Power recovery from low-grade heat by means of screw expanders

Related titles:

Fundamentals of fluid flow: Physics, numerical analysis, and measurement techniques
(ISBN 978–0–85709–477–3)

Engineering systems acquisition and support
(ISBN 978–0–85709–212–0)

Parameter identification and monitoring of mechanical systems under nonlinear vibration
(ISBN 978–1–78242–165–8)

Power recovery from low-grade heat by means of screw expanders

Ian K. Smith
Nikola Stosic
Ahmed Kovacevic

ELSEVIER

AMSTERDAM • BOSTON • CAMBRIDGE • HEIDELBERG • LONDON
NEW YORK • OXFORD • PARIS • SAN DIEGO
SAN FRANCISCO • SINGAPORE • SYDNEY • TOKYO
Woodhead Publishing is an imprint of Elsevier

WP
WOODHEAD
PUBLISHING

Woodhead Publishing is an imprint of Elsevier
80 High Street, Sawston, Cambridge, CB22 3HJ, UK
225 Wyman Street, Waltham, MA 02451, USA
Langford Lane, Kidlington, OX5 1GB, UK

British Library Cataloguing-in-Publication Data
A catalogue record for this book is available from the British Library

Library of Congress Control Number: 2013957390

ISBN 978-1-78242-189-4 (print)
ISBN 978-1-78242-190-0 (online)

For information on all Woodhead Publishing publications visit our website at http://store.elsevier.com/

Typeset by RefineCatch Limited, Bungay, Suffolk

Contents

List of figures

Preface

The US Department of Energy Annual Energy Survey for 2006 revealed that more energy is available from heat rejected by US industry than from all renewable sources combined. In the absence of any other use for such heat, it could thus be used to generate electrical power. The potential of waste and other low-grade heat sources for power production has, in fact, been known and investigated for decades, but hitherto the high cost of the equipment needed to generate power from such sources has made it economically unattractive for all but a very limited number of applications.

The main reasons for the high cost of power recovery from low-grade heat are the low conversion efficiencies possible from such sources, typical values being of the order of only 10% or less, and the relatively small amounts of thermal capacity of the majority of such heat sources, which lead to small power outputs, thereby losing the cost advantages of large-scale plant.

Owing to rising fuel prices, concerns over environmental issues and carbon taxation, and advances in engineering science, both interest in and the prospects for economically viable power recovery from such heat sources have increased. The cost of the expander, traditionally a turbine, still remains an obstacle, especially in smaller units where the number of potential heat sources is greatest.

The use of twin screw expanders instead of turbines offers a number of potential advantages in that, for some applications, they can be manufactured relatively cheaply by modifying mass-produced compressors and, because they can accept two-phase liquid–vapour mixtures, or even pure liquids, to the expansion process, it is possible to consider more efficient thermodynamic cycles than those traditionally used for turbines.

This monograph is a supplement to the authors' earlier works on screw compressors. However, those were intended primarily for the compressor industry, with the aim of improving the design and manufacture of such machines. Here, the aim is to introduce twin screw expanders to the power generation industry, where such machines are largely unknown. We therefore include less detail on the screw machine itself, much of which can be obtained from these earlier works. In addition, we have not included every type of system currently considered for power recovery from low-grade heat sources. However, we have tried to assemble in one volume an understanding of the basic essentials of the thermodynamic cycles most suited for power recovery from low-grade heat sources and how these can be enhanced by using screw expanders in place of turbines, the basic design of such screw expanders, the selection of the most suitable working fluids for use in these systems, and the principles of system optimisation.

We hope that this book will help to promote the development of cost-effective systems for power outputs in the 25–500 kW range, and thus contribute to the more efficient use of energy resources.

Acknowledgements

The authors wish to thank both their former and present colleagues, whose contributions were invaluable in the preparation of this work. In particular, Dr Roger Owen, whose monumental literature review went back as far as 1830, Dr Ruis Pitanga Marques for his work on fluid mixtures, Drs Elvedin Mujic and Ashvin Dhunput, who prepared the majority of the diagrams, and Dr Matthew Read for general assistance and suggestions.

Notation

a	speed of sound
A	area
c_p	constant pressure specific heat capacity
h	fluid specific enthalpy
H	enthalpy
h	heat transfer coefficient
m	fluid mass in chamber
$\dot{m}$	mass flow
p	fluid pressure in chamber
P	power
Q	heat transferred
q	heat transferred per unit mass
r	rotor profile radius
R	gas constant
s	specific entropy
S	entropy
t	time
T	temperature
u	specific internal energy
U	internal energy
v	fluid specific volume
V	fluid volume in chamber
w	fluid velocity
W	work
w	work per unit mass
WR	work ratio

x	optimisation variable, quality
γ	ratio of ideal gas specific heat capacities
η	efficiency
ω	angular velocity
ρ	fluid density
θ	rotor shaft angle
ζ	resistance coefficient, also exergy function

Subscripts

Comp	compressor
e	environmental
Exp	expander
in	inlet
oil	oil
ind	indicated
f	liquid
g	vapour, gain
l	leakage, loss
o	ambient conditions
out	outlet
p	previous
1	upstream
2	downstream
1,2,3,4	rotor radius index

About the authors

Ian Smith has a BSc in Mechanical Engineering from the University of Cape Town, a Diploma in Gas Turbine Technology from Imperial College, London and a PhD in Mechanical Engineering from the University of London. After three years in the aircraft engine industry at Armstrong Siddeley Motors and Lucas Gas Turbines, he joined City University, London as a Lecturer in Mechanical Engineering, subsequently being promoted to Senior Lecturer, Reader and Professor of Applied Thermodynamics. Following a year's leave of absence to work on a solar-operated Organic Rankine Cycle system at the National Physical Laboratory of Israel, in 1967, his research interests have been mainly directed at systems for the recovery of power from low-grade heat sources and working fluid properties. In 1995, he and Nikola Stosic formed a Centre for Positive Displacement Compressor Technology, with the aim of establishing a self-funding group, relying on income, almost exclusively, from industrial grants and contracts. In 1998, Ahmed Kovacevic joined Ian and Nikola. Together, their joint efforts led to international recognition of City University as a centre of excellence for the design of twin-screw machines for a wide range of applications.

Professor Smith has authored or co-authored over 170 peer-reviewed journal and conference papers, eight patents and two monographs. As a result of his work, he has received 12 professional awards and prizes.

Nikola Stosic graduated in mechanical engineering at the University of Sarajevo. He then obtained an MSc at the University of Zagreb followed by a doctorate at the University of Sarajevo, where he continued as a member of staff. There, as a result of successive promotions, he became Professor of Applied Thermodynamics, specialising in steam boilers. He began working with Ian Smith at City University in 1992, and in 1995 he was awarded a Royal Academy of Engineering chair in Positive Displacement Compressor Technology, thus enabling the two of them to establish the Centre for Positive Displacement Compressor Technology. Since 2001, Professor Stosic has been Professor of Positive Displacement Compressor Technology at City University, London.

Professor Stosic has carried out research work on a range of applications of heat and fluid flow, including boilers, furnaces and compressors. Arising from this work, he is the author and co-author of over 230 peer-reviewed publications and more than 200 major industrial reports, two textbooks on screw compressors, a textbook on computers in engineering which ran to three impressions in 4 years, and a major text on boilers and furnaces with two editions in 5 years.

In addition to gaining student silver and gold medals, Professor Stosic has received 11 professional awards and prizes.

Ahmed Kovacevic is Professor of Engineering Design at City University, London. He graduated from the University of Sarajevo, obtained an MSc at the University of Tuzla and a PhD at City University, London. He has been involved in the modelling, design and use of positive displacement screw machines for over 27 years. Since joining City University in 1998, he has carried out extensive research on CFD modelling and parametric design of twin-screw machines. Since 2008 he has held the Howden Chair in Engineering Design and Compressor Technology. Professor Kovacevic is a board member of the IMechE Fluid Machinery group. He is involved in organising three major conferences, namely: the International Compressors Conferences at City University, London, the IMechE Fluid Machinery Congress, and the Engineering and the Design Society and IED Product Design Educational Conferences.

As a result of Professor Kovacevic's research and educational work, he has authored or co-authored four books of which two are on compressors, 130 peer-reviewed journal and conference publications, and more than 50 industrial reports. He holds rights to several patents in the area of positive displacement machines, and has received seven professional awards and prizes and two teaching awards.

Introduction: power from low-grade heat

The use of heat sources at relatively low temperatures to generate mechanical power is only feasible when there is no other direct use for them. When assessing such heat sources for power generation, the main criteria for optimum design differ from those associated with power conversion from the direct combustion of fossil fuels or equivalent high-temperature sources.

First, there are no direct fuel costs associated with such power plant operation, but due to the heat being received from external heat sources, rather than internal combustion, and the lower attainable efficiencies of power plant operating over a smaller temperature range, the capital cost of such systems can be more than an order of magnitude larger than that of internal combustion power plant. Accordingly, there are two main considerations for building and installing them. These are that the capital cost per unit of net output is minimised and that the potential annual running time is sufficient to produce an economically viable payback on such large capital expenditure. In general, to meet this, both the demand for power and the supply of heat must be continuous so that there is little likelihood of significant periods of part-load operation or shut down. As fuel prices rise, the case for such systems becomes more favourable. However, the biggest economic boost to their introduction in recent years has been the legislation introduced by many governments to accelerate the development of non-fossil fuel sources and to minimise carbon dioxide production. Thus, a rational decision on the viability of a specific source of low-grade heat for power production must also include an awareness of local incentive schemes and whether power recovered from that source meets the criteria required to meet such incentive schemes.

A further feature of low-grade heat sources is that although huge amounts of heat are available globally at low temperatures, because of their distributed nature and the low thermal efficiencies of the power plant operating from them, the recoverable power output from most of them is relatively small, typically being of the order of 10–500 kW. In general, plant cost is dependent on its size raised to an exponent of 0.6–0.65 and this penalises the installation of smaller units so that until relatively recently, power outputs of less than 1 MW were rarely found to be worthwhile.

As we will show, apart from direct energy conversion devices, which are not included in this text, all systems currently considered for the recovery of power from low-grade heat are based on the pressurisation, heating, expansion, condensation and re-pressurisation of a volatile fluid in a closed cycle. The complete cycle, in some cases, may extend beyond the limits of the system such as, for example, when recovering power from geothermal resources or industrial steam. However, in all the systems described, the maximum temperatures attained are well below the metallurgical limits imposed by direct combustion systems. One consequence of this is that the thermodynamic requirements for the most suitable system are best appreciated by abandoning the traditional emphasis on the ideal Carnot cycle, the so-called Carnot efficiency and attempts, especially in steam plant, to 'Carnotise' the cycle, since these can be highly misleading.

A further aspect of power plant design, which becomes more significant in lower temperature systems, is the conflict between thermal efficiency and cost. Thus, typically, in the case of a geothermal power plant, with a brine temperature of 120 °C, the maximum temperature difference between the heat source and the coolant is only of the order of 100 °C, whereas in a fuel-fired plant it can be 20 times this value. In fact, metallurgical considerations limit this to a factor of 5–10 due to material temperature limits of 500–1000 °C, depending on the type of power plant. Consequently, the thermal efficiency of a geothermal plant may only be in the region of 10% whereas, for example, it will be 40–50% in a modern coal-fired steam plant. It follows that the condenser size per unit output of such a geothermal system will be about six times that of the coal-fired plant when operating with the same coolant conditions. The size of the condenser and cost can, of course, be reduced by increasing the temperature rise of the coolant, and thus creating a larger difference between the coolant and the condensate. However, because of the already small temperature difference between the heating medium and the coolant, this will reduce the cycle efficiency by a much larger factor than it would in a fuel-fired plant and thus significantly increase the cost per unit output of other components of the system.

Another factor, of much greater significance in recovering power from low-grade heat, is the power consumed by auxiliaries. Thus, on the assumption that the power required to operate the cooling system is of the order of 2% of the heat removed, this would constitute approximately 2% of the net power output of a modern fuel-fired system but nearly 20% of that from a system with a thermal efficiency of only 10%. Again, this could be reduced by increasing the temperature rise of the coolant, but this implies a higher condensing temperature and hence less power output from the main generator.

It follows that the balance between optimising efficiency and minimising cost in the recovery of power from low-grade heat systems is far more delicate than for fuel-fired power plant. Hence optimisation considerations become far more critical.

All these factors have been known to investigators for many years but, surprisingly, in the current rush to publish on Organic Rankine Cycle systems and the use of alternative working fluids, many present-day writers make no reference to the earlier work in this area, for which there are citations going back to the early nineteenth century. Based on the investigations of the authors and their former colleagues, extending over 45 years, an extensive bibliography, with emphasis on the earlier works has been included. Some of these may be studied profitably before engaging in new projects.

Expanders for power recovery

DOI: 10.1533/9781782421900.1

Abstract: This chapter begins with a classification of fluid machinery used for compression and expansion and explains how screw expanders differ from other types of positive displacement machine and why they can be more suitable than turbines for some applications. The mode of screw expander operation is described and details are given of the differences between oil-flooded and oil-free machines and how they are lubricated.

Key words: positive displacement machines, screw expanders, oil-free and oil-flooded machines, lubrication.

Thermodynamic machines for the compression and expansion of gases and vapours are the key components of the vast majority of power generation and refrigeration systems. Such machines can be broadly classified by their mode of operation as: turbomachines or those of the positive displacement type.

1.1 Turbines

Turbomachines effect pressure changes mainly due to dynamic effects, related to the change of momentum imparted

Published by Woodhead Publishing Limited, 2014

to the fluids passing through them. These are associated with the steady flow of fluids at high velocities. These high velocities lead to compact units, best suited for relatively large mass flow rates. Thus compressors and turbines of this type are most suitable for large-scale systems. Moreover, in smaller units the high rotational speeds required for efficient operation lead to expensive solutions to their coupling to electric generators required to produce power at 50–60 Hz.

Turbines can expand dry vapours, in which liquid forms during the expansion process or liquids in which vapour forms during the expansion process. However, at the point of admission to them, the working fluid may not be a two-phase mixture of liquid and vapour. This is because the impact of any liquid droplets accelerated to high velocity in the nozzles results in blade failure.

Turbines, in which the direction of fluid flow is associated with a falling pressure, are significantly different in their design to turbo-compressors, in which the fluid has to flow against an adverse pressure gradient. There is therefore no simple and inexpensive means of converting a turbo-compressor to operate efficiently as a turbine or vice versa.

1.2 Positive displacement machines

Positive displacement machines effect pressure changes by admitting a fixed mass of fluid into a working chamber where it is confined and then compressed or expanded, and from which it is finally discharged. Such machines must operate more or less intermittently. Such intermittent operation is relatively slow. Hence these machines are comparatively large. They are therefore better suited for smaller mass flow rates and low power inputs and outputs. A number of types of machine operate on this principle, such as reciprocating,

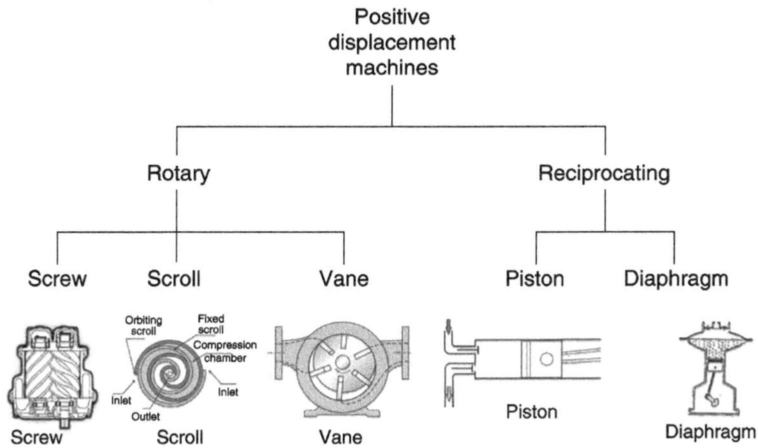

Figure 1.1 Types of positive displacement machines

vane, scroll and rotary piston machines, as shown Fig. 1.1. The modes of operation of all these types can be well understood from one- or two-dimensional representation.

1.2.1 Screw expanders

One of the most successful positive displacement machines currently in use is the twin screw, which can be used either as a compressor or an expander. An assembled view of a screw expander is shown in Fig. 1.2 with axial admission to the external high-pressure port and a radial external low-pressure discharge port.

Internally the passages are more complex, as shown in an exploded view in Fig. 1.3. The figure shows that admission to the rotors, at the high-pressure end, is mainly axial, with a small radial component of area, while discharge begins in the axial direction but is mainly radial. The flow in both directions is into a surrounding plenum chamber, leading to the external discharge port.

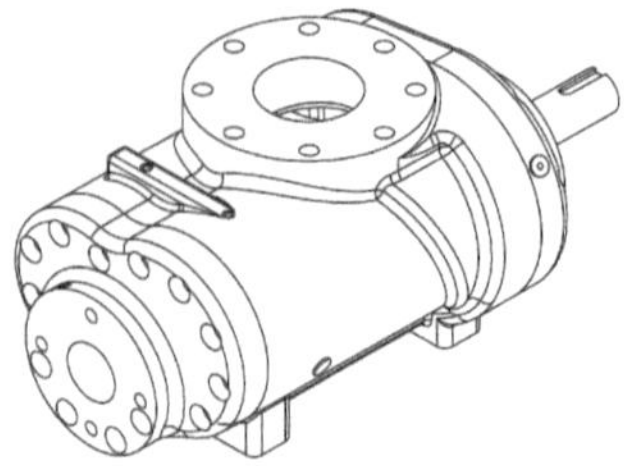

Figure 1.2 Assembled view of a screw expander

The principle of operation of screw expanders is best understood by examination of the series of diagrams in Fig. 1.4. As can be seen, it is based on volumetric changes in three dimensions rather than two. The machine consists essentially of a pair of meshing helical lobed rotors, which rotate within the surrounding casing that totally encloses them. The two meshing rotors effectively form a pair of

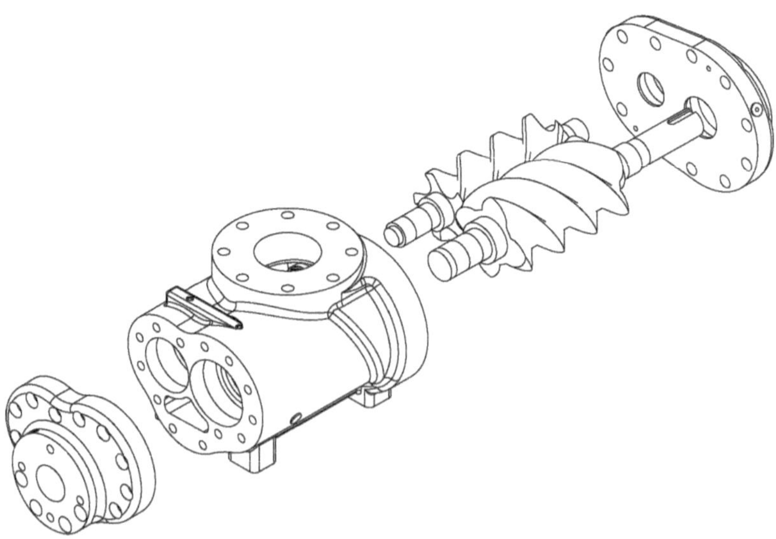

Figure 1.3 Exploded view of a screw expander

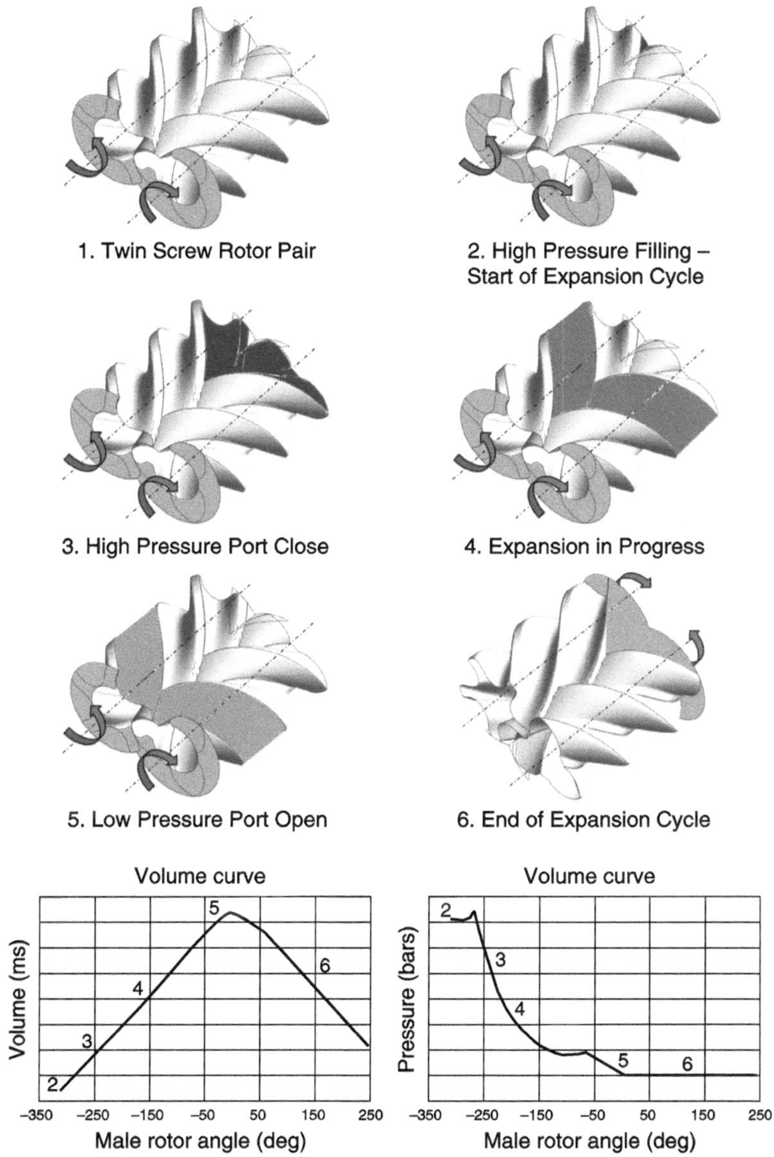

Figure 1.4 Principle of operation of a screw expander

helical gear wheels with their lobes acting as teeth. These are normally described as the male or main rotor and the female or gate rotor respectively. The space between any two successive lobes of each rotor and its surrounding casing forms a separate working chamber. Starting at the high-pressure end, as rotation proceeds, the volume of each chamber increases from zero to a value determined by the size of the inlet port. During this period of rotation, fluid flows into this space in the filling process. As rotation continues beyond this point, the working chamber is cut off from its connection with the inlet port and the line of contact between the rotors advances with a consequent increase in its volume, thus causing the trapped fluid to expand and its pressure to decrease. The volume reaches a maximum when the entire length between the lobes is unobstructed by meshing contact with the other rotor. At approximately this point, the working chamber becomes exposed to the low-pressure discharge port, at the opposite end of the rotors, and the trapped fluid begins to flow out of it. Viewed from the underside of the rotors, in the final diagram in Fig. 1.4, it can be seen that continuing rotation leads to the line of contact between the lobes receding, thus decreasing the volume of the working chamber and expulsion of the fluid, at approximately constant pressure, until the volume between the adjacent lobes is zero and all the fluid is discharged. This entire process is completed in 720 ° or two revolutions of the male rotor.

An advantageous feature of this machine is that, unlike a reciprocating or scroll expander, it has a negligible clearance volume at the high-pressure end.

The meshing action of the lobes, as they rotate, is the same as that of helical gears but, in addition, their shape must be such that at any contact position, a sealing line is formed between the rotors and between the rotors and the casing in order to minimise internal leakage between successive trapped passages. There is, however, an irreducible passage between the rotors and the cusp of the casing, which runs along the length of the expander. This is known as the blow-hole, and its shape in a given plane is illustrated in Fig. 1.5.

A further requirement is that the passages between the lobes should be as large as possible, in order to maximise the fluid displacement per revolution. Also, the contact forces between the rotors should be low so as to minimise internal friction losses.

Twin-screw machines have a number of advantages over other positive displacement types. First, unlike reciprocating machines, the moving parts all rotate and hence can run at much higher speeds. Second, unlike vane machines, the contact forces within them are low, which makes them very reliable. Third, and far less well known, unlike the reciprocating, scroll and vane machines, all the sealing lines of contact that define the boundaries of each cell chamber, are of minimum length when the pressure within the working chamber is greatest. This minimises the escape of fluid from the chamber due to leakage during the expansion process.

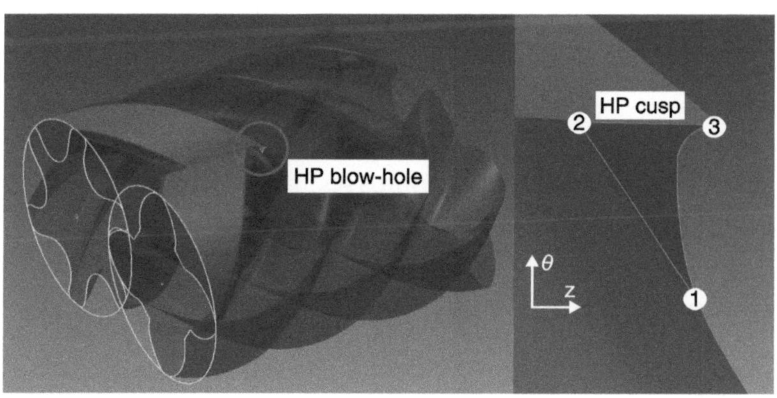

Figure 1.5 Illustration of a blow-hole

A very important feature of screw machines, whether acting as expanders or compressors, is that since the fluid velocities within them are roughly one order of magnitude less than those in turbomachines, there is little risk of damage resulting from the admission of liquid/vapour mixtures. Thus, they operate well as compressors with oil-to-gas ratios of 4:1 by mass. Similarly, as expanders they can admit fluids of any composition from pure liquid to dry vapour, while maintaining thermodynamic equilibrium between the phases. As will be shown, this enables thermodynamic cycles for power recovery to be improved. Unfortunately, other features of these machines limit their use to lower temperature systems.

Although screw machines are not commonly manufactured as expanders, they are mass-produced as compressors. For some applications, screw expanders, when required, can therefore be manufactured relatively cheaply by modifying existing compressor designs and, because of their low rotational speeds, they can easily be directly coupled to 50/60 Hz electrical generators, thereby acting as a low-cost alternative to turbines for lower power outputs.

1.2.2 Types of twin-screw machine

Twin-screw expanders and compressors may be broadly classified into two types: oil-free and oil-flooded. These are shown in Fig. 1.6, which compares machines with the same size rotors.

Oil-flooded machines

An oil-flooded machine relies on relatively large masses of oil injected with the compressed gas to lubricate the rotor motion, seal the gaps and reduce the temperature rise, when

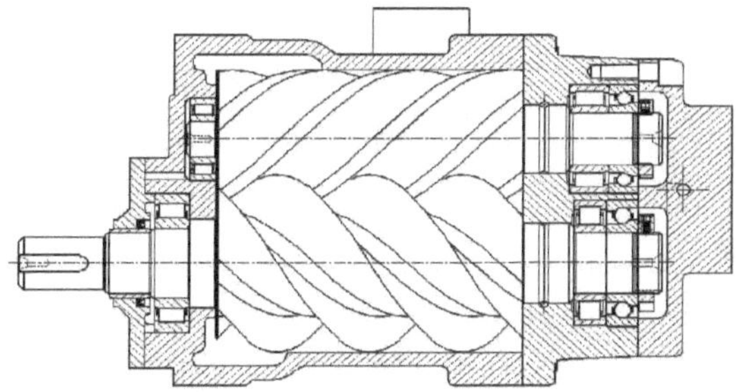

Oil-flooded

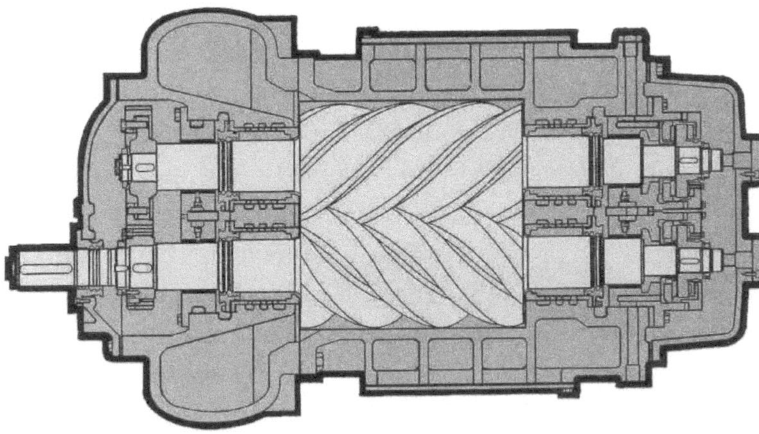

Oil-free

Figure 1.6 **Oil-flooded and oil-free compressors**

acting as a compressor. It requires an external shaft seal but no internal seals and is simple in mechanical design. Consequently, it is cheap to manufacture and highly efficient, and is widely used as a compressor in both the compressed air and refrigeration industries. Because of the oil film between the rotors, it can operate with negligible clearance between them and hence internal leakage losses are low, but the viscosity of the lubricant limits the maximum tip speeds if internal friction losses are not to become excessive. Typically, depending on the application, these vary between 20 and 50 m/s.

Oil-free machines

In an oil-free machine, there is no mixing of the working fluid with oil and contact between the rotors is prevented by timing gears which mesh outside the working chamber and are lubricated externally. In addition, to prevent lubricant entering the working chamber, internal seals are required on each shaft between the working chamber and the bearings, as well as an external shaft seal. It follows that such machines are considerably more expensive to manufacture than those that are oil-injected, but in the absence of oil they can rotate at far higher speeds, without excessive viscous drag, with typical optimum values of 80–150 m/s, depending on the application. Hence their flow capacity per unit volume is higher. However, in the absence of oil, temperature changes associated with pressure change are much greater than in oil-flooded machines and hence the clearances between the rotors have to be larger in order to avoid contact as a result of thermal expansion. Consequently, oil-free machines are generally less efficient than oil-flooded machines.

In both oil-flooded and oil-free machines, internal leakage is more or less independent of rotational speed, and decreases – as a percentage of the total flow – more or less linearly with speed, while internal friction increases with the square of the speed. This leads to a partially self-cancelling effect over a wide range of speeds and hence the efficiency of these machines varies very little over a wide speed operating range, unlike turbomachines, for which efficiencies decline rapidly away from the design speed.

Owing to the large bearing loads, resulting from changes of pressure across the rotors, internal losses in screw machines are relatively high and therefore, in most cases, they cannot attain the efficiencies of turbo-machines at the design point. However, the wide range of operating speeds over which the efficiency of screw machines varies very little, makes them more flexible and can lead to overall performance advantages over turbomachines, when operation over a wide range of conditions is required.

Lubrication systems

Both oil-flooded and oil-free machines require an external heat exchanger for the lubricating oil before it is readmitted to the machine. The oil-free machine uses this to cool the oil whether it is acting as an expander or as a compressor. To complete the circuit, an oil tank, filters and a circulating pump are required to return the oil to the bearings and timing gear, as shown in Fig. 1.7.

The oil-injected machine requires a separator to remove the oil from the discharged gas. The separated oil then has to pass through a heat exchanger, the function of which depends

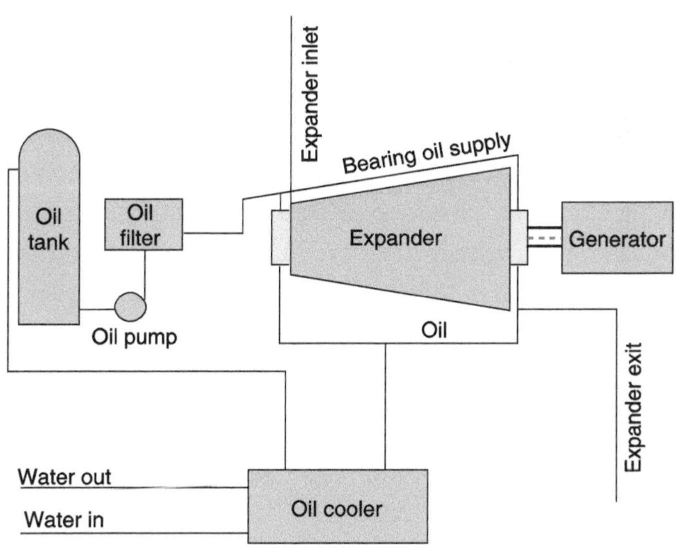

Figure 1.7 Oil-free compressor/expander lubrication system

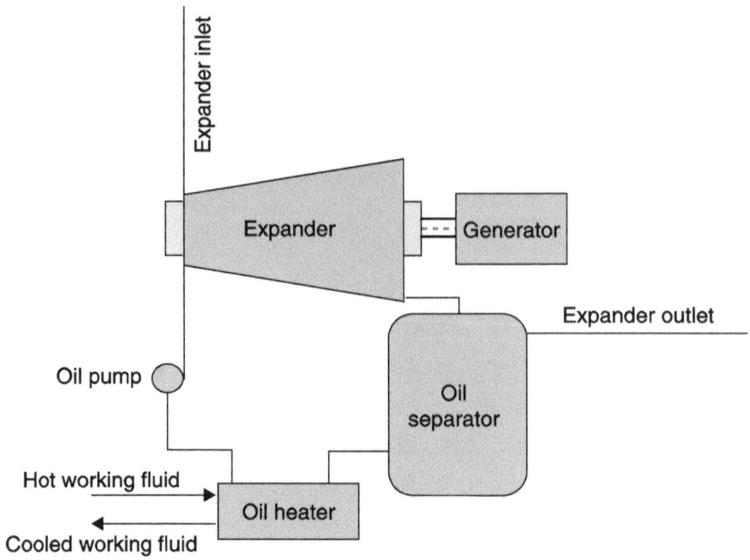

Figure 1.8 **Oil-injected expander lubrication system**

on the mode of operation. In a compressor, the oil, at high pressure, has to be cooled, and is then re-injected directly into the low-pressure region of the working chamber. However, as shown in Fig. 1.8, in an expander, the oil has to be re-pressurised in a pump and the heat exchanger has to heat it before it is returned to the high-pressure region of the working chamber. This is to prevent chilling of the hot gas or vapour entering, which would reduce the expander efficiency.

The lubrication system increases the total cost of both types of expander but the add-on cost is greater for the oil-free compressor. Thus, in the case of an oil-free expander, the total cost can be nearly an order of magnitude more than that of a bare oil-injected machine of equal size.

2

Power plant thermodynamics

DOI: 10.1533/9781782421900.13

Abstract: This chapter begins with the derivation of expressions for ideal power plant efficiencies when heat is received from a finite heat source and emphasises the importance of Carnot's principle rather than the Carnot cycle. It is shown that it is impossible to maximise power recovery and cycle efficiency simultaneously and that, as a first approximation, the atmosphere can be treated as an infinite heat sink. The importance of work ratio is emphasised and it is shown that, as a consequence, power plant cycles based on ideal gases as working fluids are unsuitable for systems recovering power from low-grade heat. When generating power from an external heat source, the importance of matching the temperature changes of the heat source and the working fluid in the primary heat exchanger is demonstrated, and it is shown that this limits the value of Carnot or Stirling cycles. Cycles based on pressurising, evaporating, expanding and condensing volatile fluids are shown to be superior both because of their very high work ratios and their superior matching characteristics. The principles of working fluid selection are described for selecting volatile fluids that combine good temperature matching and high work ratio in order to maximise the power recovery from single-phase heat sources.

Published by Woodhead Publishing Limited, 2014

Key words: maximum work, ideal cycle efficiencies, work ratio, matching, heat recovery and conversion efficiencies, practical cycles, organic fluids, working fluid selection.

2.1 Maximum work

The main sources of energy for thermal power plant heat are derived from flowing fluids. These can initially be cool, such as air and fuel entering a steam power plant boiler or an internal combustion engine or, in the case of low-grade heat, hot, such as internal combustion engine exhaust gases, waste heat streams from chemical and industrial processes and geothermal brines. Analysis of energy recovery from them must therefore be based on the assumption of steady or quasi-steady flow from a finite heat source; that is, transfer of heat from them, either as a result of combustion or by direct heat transfer, causes their temperature to change. Hence to evaluate their potential for conversion to power, more generalised expressions are required for maximum work and efficiency than those derived from consideration of an infinite static heat source, on which the Carnot cycle is based.

Consider the steady flow of fluids through a control volume, surrounding a prime mover of any type, such as an internal combustion engine, as shown in Fig. 2.1. This is

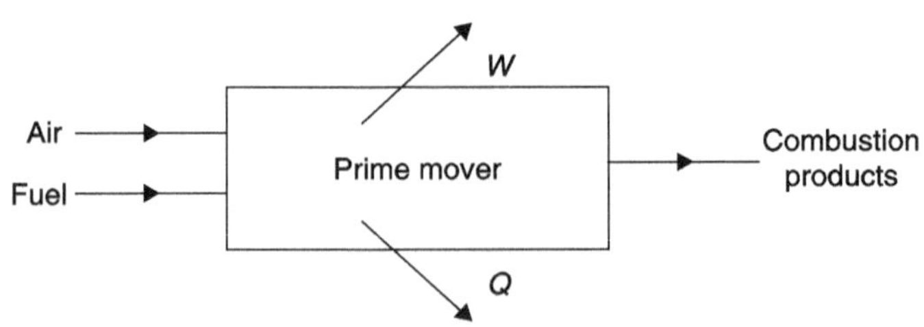

Figure 2.1 Internal combustion engine

located in an environment at temperature T_o to which heat may be rejected. The environment is usually the atmosphere, although sometimes it may be a supply of cooling water from a river or sea. The environment may be considered to be large enough for T_o not to change as a result of heat transfer to it.

Under steady flow conditions, the equation of continuity may be expressed as:

$$\dot{m}_{in} = \dot{m}_{out}$$

Neglecting potential and kinetic energy effects, the steady flow energy equation may be written as:

$$(-\dot{Q}) - \dot{W} = \dot{m}_{out} h_{out} - \dot{m}_{in} h_{in} = \dot{m}(h_{out} - h_{in})$$
$$\therefore \quad \dot{W} = \dot{m}(h_{in} - h_{out}) - \dot{Q}$$

As a result of the heat transfer to the environment, its entropy increases by ΔS_e, since:

$$\Delta \dot{S}_e \equiv \frac{\dot{Q}}{T_o} \quad \therefore \dot{Q} = T_o \Delta \dot{S}_e$$
$$\therefore \quad \dot{W} = \dot{m}(h_{in} - h_{out}) - T_o \Delta \dot{S}_e$$

If the fluids, the control volume and the environment are considered jointly as a single isolated system, then from Clausius's inequality, its total rate of change of entropy is:

$$\dot{m}(s_{out} - s_{in}) + \Delta \dot{S}_e \geq 0$$
$$\therefore \quad \Delta \dot{S}_e \geq \dot{m}(s_{in} - s_{out})$$
$$\therefore \quad \dot{W} \leq \dot{m}[(h_{in} - h_{out}) - T_o(s_{in} - s_{out})]$$

Hence, if all the processes are reversible, the maximum power output can be expressed as:

$$\dot{W}_{\max} = \dot{m}[(h - T_o s)_{in} - (h - T_o s)_{out}]$$

Or, expressed in terms of unit mass flow:

$$\frac{\dot{W}_{max}}{\dot{m}} = w_{max} = (h - T_o s)_{in} - (h - T_o s)_{out} \qquad [2.1]$$

The term $h - T_o s$ is known as the exergy function ζ, or exergy. Changes in exergy can include the effects of chemical reactions and electrolytic effects. Thus equation [2.1] defines the maximum work recoverable from any system, including fuel cells, internal combustion engines and heat engines together with their source of heat.

As shown in Fig. 2.1, in the case of an internal combustion engine or a fuel cell, maximum power would be recovered when the fuel and air enter at atmospheric temperature and pressure and the combustion or reaction products leave at atmospheric temperature and pressure. Under those conditions, it follows from Eq. 2.1 that:

$$w_{max} = (h_{reactants} - h_{products})T_o - T_o(s_{reactants} - s_{products})$$
$$= [\Delta H']T_o - T_o(s_{reactants} - s_{products})$$

where $\Delta H'$ is the enthalpy of reaction or calorific value.

Define the Gibbs function: $g \equiv h - T\,s$.
Hence $w_{max} = -[\Delta g']T_o$

When hydrocarbon fuels and air react to form carbon dioxide and water, generally $s_{products} < s_{reactants}$. It therefore follows that $-[\Delta g']T_o < -[\Delta H']T_o$. However, since the additional term $T_o(s_{reactants} - s_{products})$ is generally not large, the efficiencies of thermal power plant based on combustion processes are calculated on the assumption that the maximum recoverable work is equal to the enthalpy of reaction of the fuel. This is a convenient approximation but is not strictly correct.

If Eq. 2.1 is applied to determine the maximum power recoverable from a steady flow of dry saturated steam, at temperature T_1, acting as a source of energy to a heat engine,

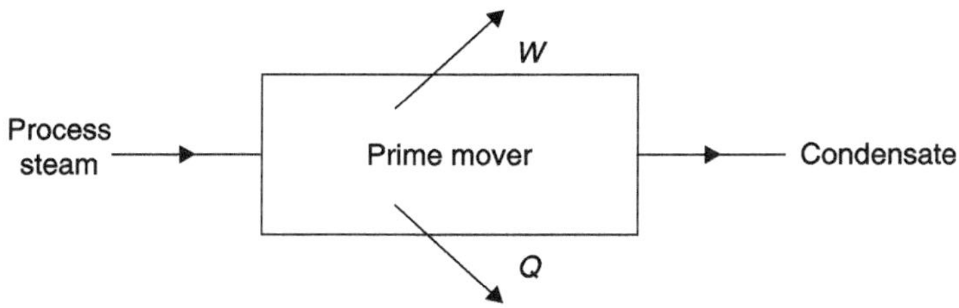

Figure 2.2 Power plant receiving heat from process steam

from which it leaves as saturated liquid at the same temperature and pressure, as shown in Fig. 2.2, then

$$w_{max} = h_{g1} - h_{f1} - T_o(s_{g1} - s_{f1}) = h_{fg1} - T_o s_{fg1}$$

In this case, the heat supplied $Q_{in} = h_{fg1}$ and for saturated steam $h_{fg} = Ts_{fg}$.

The maximum thermal efficiency of a heat engine is normally expressed in the form of a cycle efficiency, η_{cycle}, where:

$$\eta_{cycle} \equiv \frac{\text{maximum work}}{\text{heat supplied}} = \frac{T_1 s_{fg1} - T_o s_{fg1}}{T_1 s_{fg1}} = \frac{T_1 - T_o}{T_1} \quad [2.2]$$

This corresponds to the efficiency of a heat engine operating on a Carnot cycle between the steam and the environmental temperatures. The result is hardly surprising, since steam condensing at constant pressure and temperature acts as an isothermal heat source, and therefore has equivalent characteristics to that of an infinite source of heat.

Consider now the steady flow of a hot gas or liquid stream, acting as a source of heat to a heat engine, which it enters at temperature T_1 and leaves at temperature T_2, without loss of pressure, as shown in Fig. 2.3. In that case,

$$h_{in} - h_{out} = c_p(T_1 - T_2)$$

and

$$s_{in} - s_{out} = \int_2^1 \frac{dh - vdp}{T} = \int_2^1 \frac{c_p dT}{T} - R\int_2^1 \frac{dp}{p} = c_p \ln\frac{T_1}{T_2}$$

$$\therefore w_{\max} = c_p(T_1 - T_2) - T_o c_p \ln\frac{T_1}{T_2} \quad [2.3]$$

However, the heat input is $q_{in} = c_p(T_1 - T_2)$,

$\therefore$ maximum cycle efficiency

$$= \frac{\text{maximum work}}{\text{heat supplied}} = 1 - \frac{T_o \ln\left[\frac{T_1}{T_2}\right]}{T_1 - T_2} \quad [2.4]$$

This has a limiting value when the hot fluid leaves at the temperature of the environment, when:

$$\eta_{cycle} = 1 - \frac{T_o \ln\frac{T_1}{T_o}}{T_1 - T_o} \quad [2.5]$$

Equations 2.3 to 2.5 give limiting values to attainable power outputs and cycle efficiencies from hot fluid streams, which correspond to the overwhelming number of types of heat source from which power is to be recovered in low-grade heat systems. They also apply equally well to high-

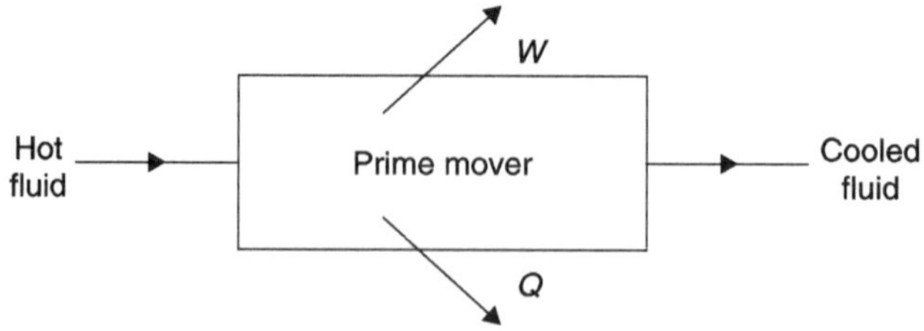

Figure 2.3 Power plant receiving heat from a hot fluid

temperature systems, such as steam power plant receiving heat from the direct combustion of fuel. However, they are rarely cited as an upper limit in thermodynamics textbooks and, as derived above, they give little guide to the practical power plant designer on what combination of ideal processes is required to obtain these values. The following alternative approach may provide better insights into what is required.

Consider a hot fluid stream, initially at temperature T_1, supplying heat to an ideal heat engine, which, in turn, rejects heat to an infinite heat sink at temperature T_0. Any heat transfer from the hot stream must be associated with a decrease in its temperature. Thus, per unit mass flow rate of fluid, at any temperature T,

$$dq = -c_p \, dT$$

For the receipt of heat dq, the engine does work dw, where:

$$dw = dq - dq_0$$

As shown in Fig. 2.4, for an ideal infinitesimal heat engine, according to the Second Law:

$$\frac{dq}{T} = \frac{dq_0}{T_0}$$

$$\therefore dw = \left[1 - \frac{T_0}{T}\right] dT = c_p \left[\frac{T_0}{T} - 1\right] dT \qquad [2.6]$$

It follows that the ideal system for recovery of power under these conditions is a sequence of infinitesimal Carnot cycles, each operating with a decreasing temperature difference between the heat source and sink, as shown in Fig. 2.5. Accordingly, the maximum power output can be obtained by integrating Eq. 2.6 between the initial fluid stream temperature, T_1, and its exit temperature from the power plant, T_2.

$$\therefore \text{maximum work} = \int_1^2 dw = c_p \int_{T_1}^{T_2} \left[\frac{T_0}{T} - 1 \right] dT$$

$$= c_p \left[T_1 - T_2 - T_0 \ln \frac{T_1}{T_2} \right].$$

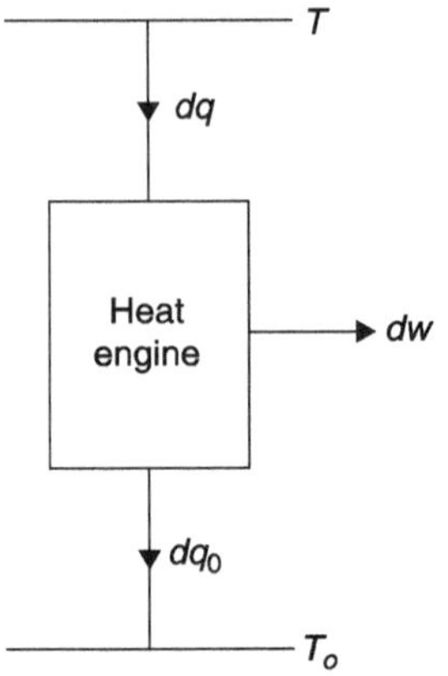

Figure 2.4 Infinitesimal heat engine

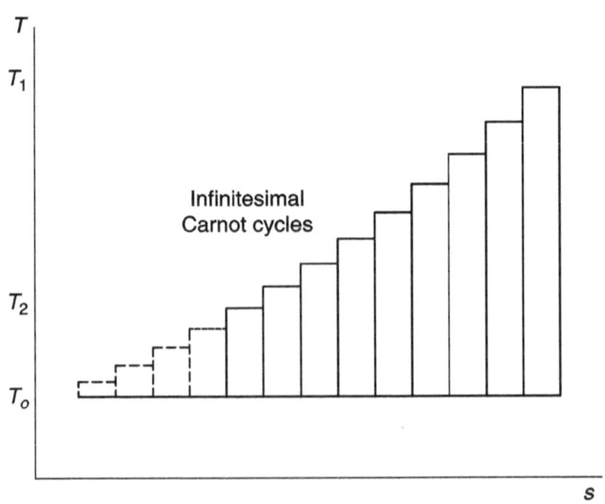

Figure 2.5 Temperature–entropy diagram for ideal recovery of power from a hot fluid stream

And the corresponding cycle efficiency is obtained by dividing by the heat input $c_p(T_1 - T_2)$ to obtain:

$$\eta_{cycle} = 1 - \frac{T_0 \ln \frac{T_1}{T_2}}{T_1 - T_2}$$

This result coincides with Eq. 2.4. It follows that the ideal cycles for power recovery from a hot fluid stream can be drawn in temperature–entropy coordinates, as shown in Fig. 2.6. These have been described in previous literature as either ideal quadlateral or ideal trilateral cycles, depending on whether the hot fluid is cooled down to some intermediate temperature T_2 or to the ambient temperature T_o, respectively.

Taking the argument a stage further, in most cases, the supply of coolant, to remove the heat rejected from a thermal power plant, is in fact limited and, locally, the coolant temperature may rise significantly as in the case of a water- or air-cooled condenser. In that case, when considering the power recovery from the steady flow of a hot fluid at constant pressure, as considered in the last example, the control volume can be extended to include the coolant flow, as

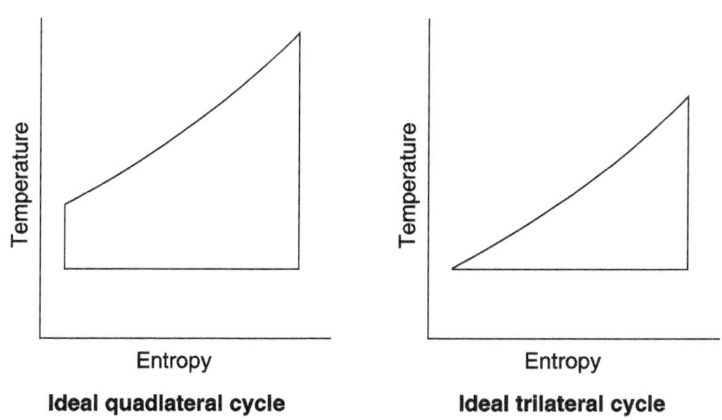

Figure 2.6 Ideal cycles between a finite heat source and an infinite heat sink

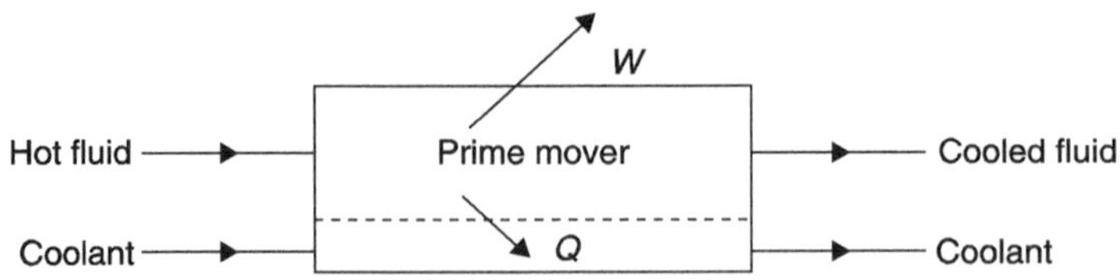

Figure 2.7 **Power plant receiving heat from and rejecting to external fluid streams**

shown in Fig. 2.7, thereby eliminating any external heat transfer outside the control volume boundaries.

If the coolant enters at the temperature of the environment, T_o, and leaves at some higher temperature, T_3, then, assuming that the fluid flow rate of the hot fluid is m_h and its specific heat capacity at constant pressure is c_{ph} while the corresponding properties of the coolant are m_c and c_{pc} respectively, then:

$$\begin{aligned}\text{maximum work} = &\, m_h[h_1 - h_2 - T_o\,(s_1 - s_2)] \\ &+ m_c[h_o - h_3 - T_o(s_o - s_3)]\end{aligned}$$

If both the hot fluid and the coolant flow without change in pressure, then:

$$\begin{aligned}\text{maximum work} = &\, m_h\left[c_{ph}(T_1 - T_2) - T_o c_{ph} \ln\frac{T_1}{T_2}\right] \\ &+ m_c\left[c_{pc}(T_o - T_3) - T_o c_{pc} \ln\frac{T_o}{T_3}\right]\end{aligned}$$

As is the case for a finite heat sink, the heat input will remain as:

$$Q_{in} = m_h c_{ph}(T_1 - T_2)$$

Bearing in mind that for the power recovery and cycle efficiency to be a maximum, the net entropy production of the system, comprising the heat source, the heat engine and the coolant, must be zero, it can be shown that:

$$\text{maximum work} = m_h c_{ph}[(T_1 - T_2) - (T_3 - T_o)\frac{\ln\frac{T_1}{T_2}}{\ln\frac{T_3}{T_o}}] \qquad [2.7]$$

and

$$\text{ideal cycle efficiency} = 1 - \frac{[(T_3 - T_o)]}{(T_1 - T_2)}\frac{\ln\frac{T_1}{T_2}}{\ln\frac{T_3}{T_o}} \qquad [2.8]$$

As for the previous case, these values will be a maximum when all the heat is extracted from the hot fluid and it leaves the power plant with $T_2 = T_o$, when:

$$\text{ideal cycle efficiency} = \eta_{cycle} = 1 - \frac{[(T_3 - T_o)]}{(T_1 - T_0)}\frac{\ln\frac{T_1}{T_0}}{\ln\frac{T_3}{T_o}}. \qquad [2.9]$$

The ideal cycles in this case are slightly different, as shown in Fig. 2.8, assuming that the temperature rise of the coolant is

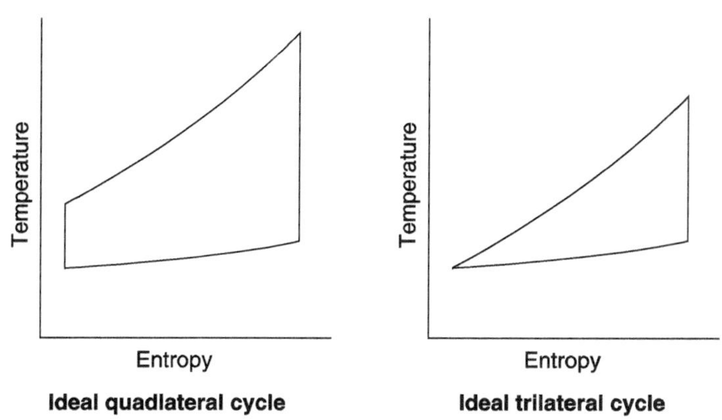

Figure 2.8 Ideal cycles between a finite heat source and a finite heat sink

relatively small, compared with the fall in temperature of the heating medium.

2.2 Some observations

Figure 2.9 compares cycle efficiencies assuming a Carnot cycle, derived from Eq. 2.2, and an ideal trilateral cycle, derived from Eqs. 2.4 and 2.7, assuming an ambient temperature of 15 °C.

First, as can be seen, efficiencies predicted from the assumption of a Carnot cycle, as the ideal, are nearly twice the real ideal values, as derived from the trilateral cycle. Clearly, the Carnot cycle cannot be used as a basis for comparing real efficiencies with what is theoretically possible,

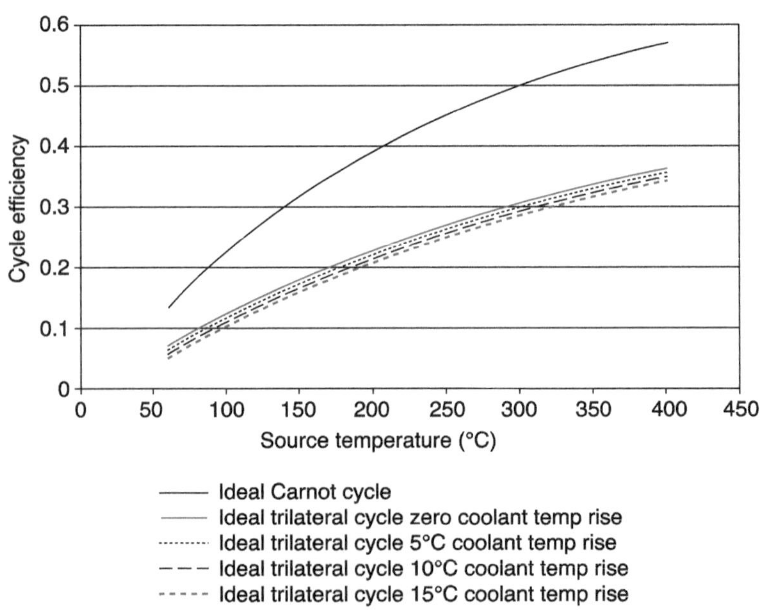

Figure 2.9 **Comparison of Carnot cycle with ideal trilateral cycle**

when considering finite sources of heat. In fact, there are very few heat sources that can be considered to be infinite and the Carnot cycle is better taken as a concept that is used for two main purposes, namely:

1. The establishment of Carnot's principle, i.e. for maximum efficiency, heat should be input to a thermal power plant system at the highest possible temperature and rejected from it at the lowest possible temperature.
2. The derivation of the corollaries of the Second Law of Thermodynamics, rather than the basis for practical heat engine design.

Second, it can be seen that except when considering power recovery from very low-temperature heat sources, the effect of the temperature rise of the coolant on the cycle efficiency is not very great and the assumption of an infinite heat sink, as used to derive Eqs. 2.3 to 2.5, in order to estimate maximum power recovery and cycle efficiency, is a good approximation.

Figure 2.10 shows another aspect of ideal power recovery from a finite heat source. Typically, in this case, a hot gas stream initially at 350 °C is used as a source of heat from which power recovery and cycle efficiency have been estimated by the solution of Eqs. 2.3 and 2.4. As can be seen, the power recoverable per unit mass flow of the gas increases as the gas temperature leaving the power plant approaches the ambient temperature. However, the greater power output is achieved with a reduced cycle efficiency. The reason for this is that the additional power generated is derived from a stream of steadily decreasing temperature. Hence, the efficiency with which the additional power is recovered declines as the exhaust gas exit temperature falls.

In practice, there are often practical limitations to the minimum attainable temperature. Typically, in the case of engine exhaust gases, especially with fuels containing sulphur,

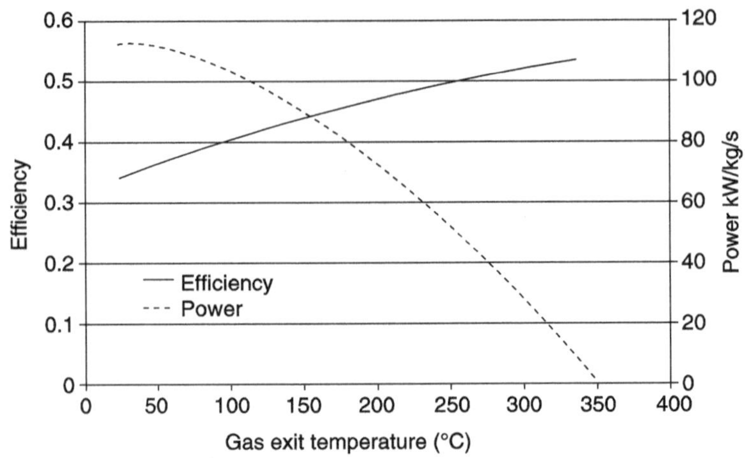

Figure 2.10 The effect of the fluid exit temperature on ideal efficiency and power output. Ideal performance: gas inlet temperature 350 °C, ambient temperature 15 °C

the exit temperature must be above that of the acid dewpoint, in order to avoid condensation, while in the case of geothermal brines, it must be high enough to avoid the precipitation of any dissolved solids, such as silicates and salts, which block the heat exchangers.

In general, the higher the cycle efficiency, the less heat transfer is needed per unit power output. Hence, the size and cost of the heat exchangers are reduced and vice versa. Thus, the power plant designer has to make a compromise between maximising power recovery and minimising cost per unit output.

2.3 Practical considerations

2.3.1 Work ratio

A much-neglected aspect of thermal power plant theory is the concept of work ratio. However, its consideration gives clear guidelines to which types of system are the most suitable for power recovery from low-grade heat sources.

Work ratio, *WR*, is defined as:

net work output in cycle ÷ positive work input in cycle

It is most easily visualised from the consideration of pressure–volume diagrams. Two cases should be included. When a cycle is comprised of a sequence of non-flow processes, such as an Otto or Joule cycle, then, as is well known from elementary thermodynamic theory, the area under a pressure–volume diagram, derived from the $\int pdV$, is equal to the net work carried out in the cycle. Similarly, when the cycle is comprised of a succession of steady flow processes, such as in a turbine-driven, condensing steam power plant, then the net work carried out in the cycle is = $-\int Vdp$.

Arbitrary examples of these are shown in Fig. 2.11, where the positive, negative and net work carried out in the cycle can be clearly identified. As can be seen, effectively, the pressure–volume diagram illustrates work ratio in the same way that the temperature–entropy diagram illustrates the

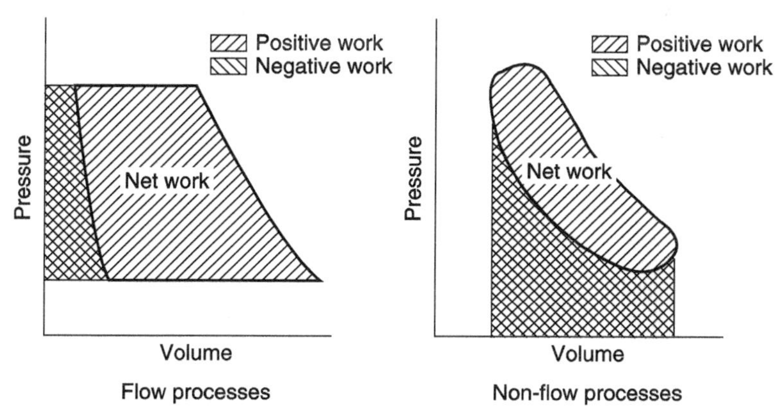

Figure 2.11 Pressure–volume diagrams for power plant cycles based on flow and non-flow processes

cycle efficiency. In both cases, the desirable aim is to have a large enclosed area, compared with that of the total gross area enclosed under the appropriate coordinate.

As an aside, it is worth noting that the role of the pressure–volume diagram is often confusing because work as expressed as the $\int pdV$ refers to pressure–volume changes in a fixed system, whereas in air compressors, steam engines and internal combustion engines, there is an intermittent inflow and outflow of fluid at different stages in the 'cycle' of processes, yet nonetheless the cyclic work is accurately defined by such a diagram. It should therefore be stated explicitly that this assumption is correct for any system that uses a positive displacement machine such as a piston and cylinder, or screw or scroll machine as the prime mover. This is because the machine itself, in such cases, should be regarded as the system on which the forces, induced by pressure change, act. Put differently, in a closed cycle of processes, $\oint pdV = -\oint Vdp$. Hence, the relationship is correct whether the sequence of processes describes flow or non-flow processes. However, the thermodynamic processes of the working fluid cannot be correctly inferred from this. Thus, in the case of a petrol engine, allegedly operating on the Otto cycle, on the opening of the exhaust valve, the combustion products undergo a sudden expansion and not a constant-pressure volume change.

The significance of work ratio can be appreciated from the following simplified analysis. First, for convenience, we may define component efficiencies in general terms. *Expansion efficiency* can be defined as η_{exp}, where:

$$\eta_{exp} \equiv \frac{\text{work done}}{\text{work done in the absence of all irreversibilities}}$$

Similarly, *compression efficiency* can be defined as η_{comp}, where:

$$\eta_{comp} \equiv \frac{\text{work input in the absence of all irreversibilities}}{\text{actual work input}}.$$

These definitions can be applied to both isentropic and isothermal processes within a cycle.

Consider any power plant cycle. Since the net work output, W_{net}, must be the difference between the total expansion work, W_{exp}, minus the total compression work, W_{comp}, it follows that:

$$W_{exp} = \frac{W_{net}}{WR} \quad \text{and} \quad W_{comp} = W_{net}\left(\frac{1}{WR} - 1\right)$$

Hence, for any ideal cycle:

$$W_{net\,ideal} = \frac{W_{net\,ideal}}{WR} - W_{net\,ideal}\left(\frac{1}{WR} - 1\right)$$

If the definitions for expander and compressor efficiency are now included to estimate the achievable net work output in a practical cycle, then:

$$\begin{aligned} W_{net} &= \frac{W_{net\,ideal}}{WR}\eta_{exp} - \frac{W_{net\,ideal}}{\eta_{comp}}\left(\frac{1}{WR} - 1\right) \\ &= W_{net\,ideal}\left(\frac{\eta_{exp}}{WR} - \frac{1 - WR}{\eta_{comp} WR}\right) \end{aligned} \qquad [2.10]$$

Since the cycle efficiency, η_{cycle}, is defined as W_{net}/Q_{in}, then it follows that:

$$\eta_{cycle} = \eta_{ideal\,cycle}\left(\frac{\eta_{exp}}{WR} - \frac{1 - WR}{\eta_{comp} WR}\right)\frac{Q_{in\,ideal}}{Q_{in}}$$

If we make the simplifying assumption that the heat input to an ideal cycle, $Q_{in\,ideal}$, is identical to the heat input to the real cycle, in which the expansion and compression processes are irreversible, then:

$$\frac{\eta_{cycle}}{\eta_{ideal\ cycle}} = \left(\frac{\eta_{exp}}{WR} - \frac{1-WR}{\eta_{comp} WR}\right) \qquad [2.11]$$

In reality, the heat input to a given cycle operating between specified temperature limits will vary with the component efficiencies. However, in most modern thermal power plant, where these are high, the error resulting from this assumption will be of a secondary order and will not be sufficiently large to affect the use of Eq. 2.11 to evaluate the effect of the work ratio on practical cycle efficiency. This is shown in Fig. 2.12, which was obtained by solving Eq. 2.11 for a variety of assumed compressor and expander efficiencies.

It is clear that when the work ratio approaches unity, the achievable cycle efficiency is not greatly reduced by the irreversibility of the compression or pressurisation and expansion processes. However, when the work ratio is low, the practically attainable cycle efficiency may be only a small fraction of the theoretical ideal value and may even result in no net power output. Its significance is therefore especially

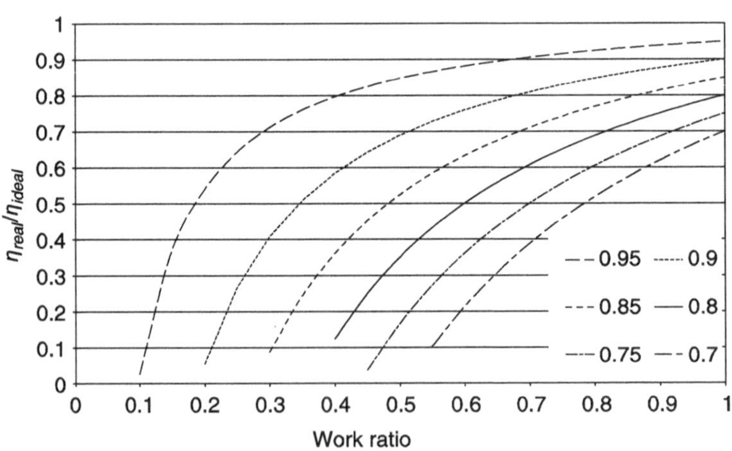

Figure 2.12 **The effect of work ratio on practical cycle efficiency**

important when dealing with systems for recovery of power from low-grade heat, where even ideal cycle efficiencies are lower than those normally achieved from direct combustion of fuel or higher temperature heat sources.

The usefulness of the work ratio concept can be illustrated by considering ideal trilateral and quadlateral cycles as 'air standard' cycles, as used in engineering thermodynamics textbooks, with an ideal gas as the working fluid. The ideal trilateral cycle can be reproduced exactly by a sequence of processes comprising constant pressure heating, isentropic expansion and isothermal recompression. These are shown in Fig. 2.13 on both $T-s$ and $p-V$ coordinates. In this case, it follows from Eq. 2.3 that the positive work $= c_p(T_1 - T_o)$ and the negative work $= c_p T_o \ln(T_1 - T_o)$. It therefore follows that for the ideal trilateral cycle using a perfect gas as the working fluid:

$$\text{work ratio} = 1 - \frac{T_0 \ln \dfrac{T_1}{T_0}}{T_1 - T_0} = \text{cycle efficiency} \qquad [2.12]$$

Clearly, at low temperatures, where the cycle efficiency is low, the work ratio will also be low. Assuming, rather optimistically, that the system will have an expander with an

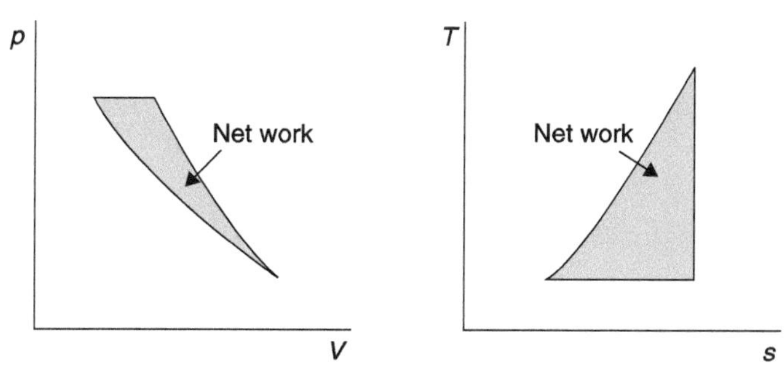

Figure 2.13 The ideal trilateral cycle using a perfect gas

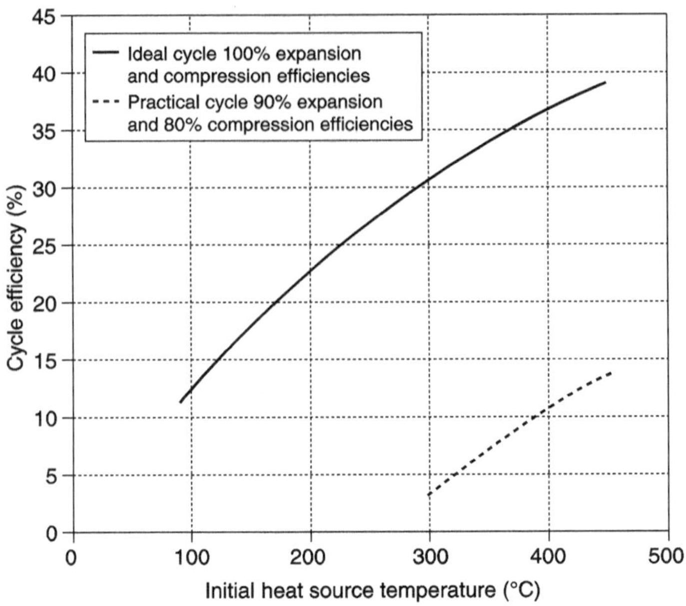

Figure 2.14 **Comparison of ideal and achievable trilateral cycle efficiency using a perfect gas: heat rejection temperature 15 °C**

adiabatic efficiency of 90% and an isothermal compression efficiency of 80%, then the achievable cycle efficiency can be seen compared with the ideal, based on the solution of Eqs. 2.11 and 2.12, in Fig. 2.14.

Similarly, the ideal quadlateral cycle, using an ideal gas as the working fluid, is shown both in $T-s$ and $p-V$ coordinates in Fig. 2.15, where:

$$\text{work ratio} = \frac{T_1 - T_2 - T_0 \ln \dfrac{T_1}{T_2}}{T_1 - T_0} \qquad [2.13]$$

Again, based on the solution of Eqs. 2.4, 2.11 and 2.13, the ideal and achievable cycle efficiencies are shown in Fig. 2.16. It is clear that in this case the need for adiabatic

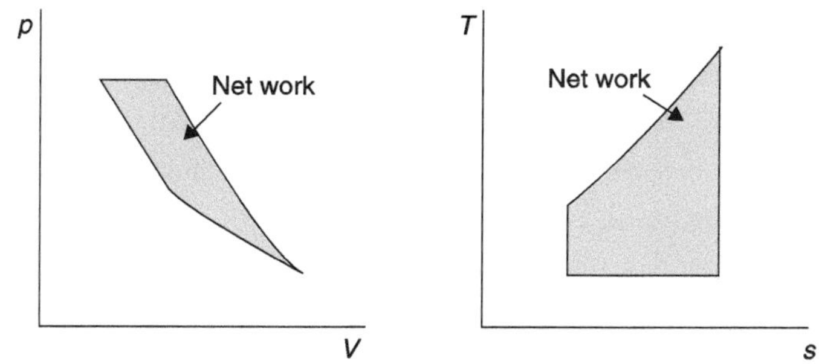

Figure 2.15 The ideal quadlateral cycle using a perfect gas

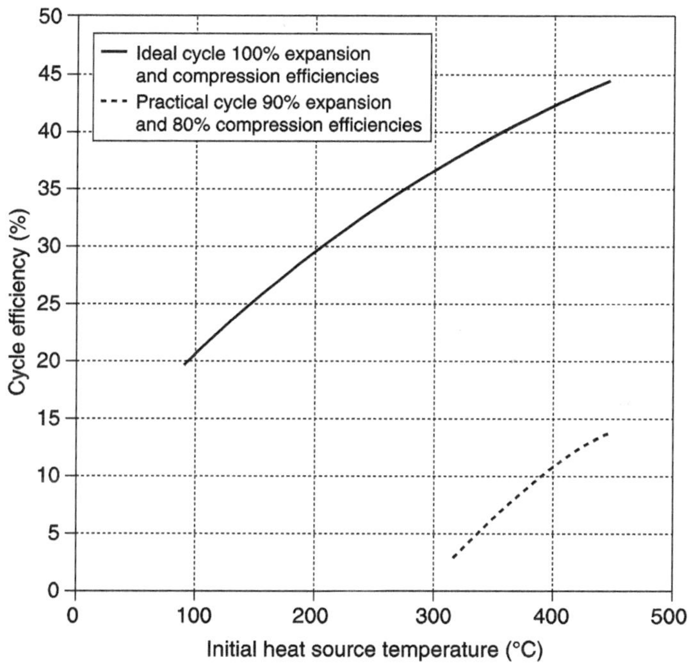

Figure 2.16 Comparison of ideal and achievable quadlateral cycle efficiencies using a perfect gas. Minimum heat source temperature 80 °C, heat rejection temperature 15 °C

compression, in addition to isothermal compression of the gas, makes this cycle even less efficient than the ideal trilateral cycle, despite its more favourable $T-s$ diagram.

It is clear that although, at first sight, such cycles may appear to be ideal, in practice the low work ratios, associated with the need to recompress the gas, make them totally unsuited for power recovery from low-grade heat sources. Preliminary evaluations of this kind can therefore save much unnecessary time and effort in estimating what systems may or may not be suitable for low-temperature applications.

2.3.2 Matching

Some heat sources, such as steam or pressurised geothermal hot water produced from natural aquifers, or industrial process steam, can be expanded directly to produce mechanical power. However, most of the low-grade heat sources available for power recovery are hot liquids or gases at near atmospheric pressure, which can only be used to supply heat to be converted to power in a heat engine operating in a closed cycle. In that case, a most important consideration is how well the temperature change of the heat source matches that of the working fluid in the heat exchanger of the power plant cycle that receives the heat. Thus, if the heat source and sink are included in a temperature–entropy diagram, the ideal combination of source, sink and power plant cycle should appear as shown in Fig. 2.17, where some allowance has to be made for a temperature difference between the heat source and the heat engine converting the heat received into power as well as between the heat engine and the coolant removing the rejected heat.

The magnitude of these temperature differences is determined by a compromise between minimising their value, in order to maximise the efficiency of the heat engine

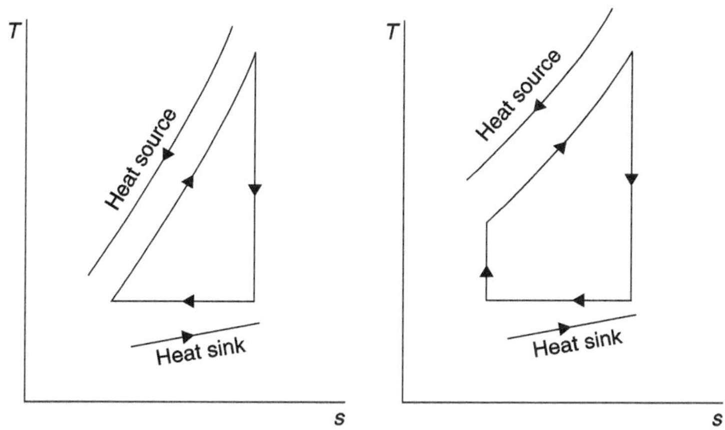

Figure 2.17 Ideal trilateral and quadlateral cycles matched to the heat source and sink

cycle and the need to minimise the heat exchanger costs, which must increase, as the temperature difference across the heat exchanger is decreased. This factor is far more critical when recovering power from lower-temperature sources than in high-temperature systems receiving heat from direct combustion products, where, in many cases, the upper temperature in the heat engine cycle is limited by metallurgical considerations, leaving a large temperature difference for heat transfer.

It has already been shown that cycles based on ideal gases that replicate ideal trilateral and quadlateral cycles are impracticable. An alternative is to consider cycles in which the heating and cooling processes do not follow the ideal requirement so closely but which are inherently more efficient. As an example, consider a Stirling cycle engine as a means of power recovery. This has been chosen because it is thermodynamically equivalent to a Carnot cycle and therefore of potentially high efficiency.

As shown in Fig. 2.18, the Stirling cycle combines isothermal expansion and compression with constant volume

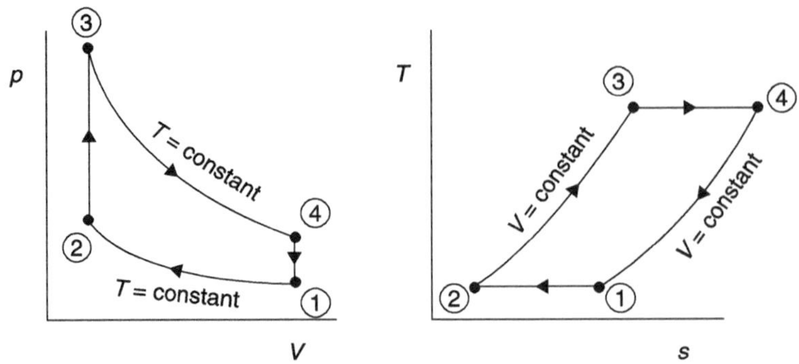

Figure 2.18 **Ideal Stirling cycle on *p*–*V* and *T*–*s* coordinates**

compression and expansion, achieve by internal heat transfer between the expanding and compressing gases. It can be shown that:

$$\eta_{cycle} = \frac{T_{\max} - T_{\min}}{T_{\max}} = \text{work ratio} \qquad [2.14]$$

When using such a cycle to recover heat from a finite heat source, the problem is to find the optimum value for $T_{\max}$ in order to maximise the power generated.

Consider the cooling of a hot fluid stream, as shown in Fig. 2.19. Given that the fluid is initially at temperature T_1, the maximum heat recovery is obtained when it is cooled to the environmental temperature T_o. When cooled to some intermediate temperature, T, the amount of heat, Q, is recovered with a heat recovery efficiency $\eta_{heat\ recovery}$ that can be defined as:

$$\eta_{heat\ recovery} \equiv \frac{T_1 - T}{T_1 - T_0} = \frac{Q}{Q_{max}} \qquad [2.15]$$

To obtain the maximum power output, we need to optimise the conversion efficiency $\eta_{conversion}$, where:

$$\eta_{conversion} \equiv \frac{W}{Q_{max}} = \frac{W}{Q}\frac{Q}{Q_{max}} = \eta_{cycle}\,\eta_{heat\ recovery} \qquad [2.16]$$

Assuming that heat rejection from the engine can be at the environmental temperature T_o, then:

$$\eta_{conversion} = \frac{T - T_0}{T}\frac{T_1 - T}{T_1 - T_0}$$

By differentiating with respect to T, it can be shown that $\eta_{conversion}$ is a maximum when $T = \sqrt{T_1 T_0}$. In which case:

$$\eta_{conversion} = \frac{\sqrt{T_1 T_0} - T_0}{\sqrt{T_1 T_0}}\frac{T_1 - \sqrt{T_1 T_0}}{T_1 - T_0} \qquad [2.17]$$

For purposes of illustration, the thermodynamic processes involved in the cycle, the heat transfer medium and the coolant are shown in Fig. 2.20(a), where the loss in potential work, due to the constant temperature heating process in the Stirling cycle, is shown by the shaded area. This, of course, is

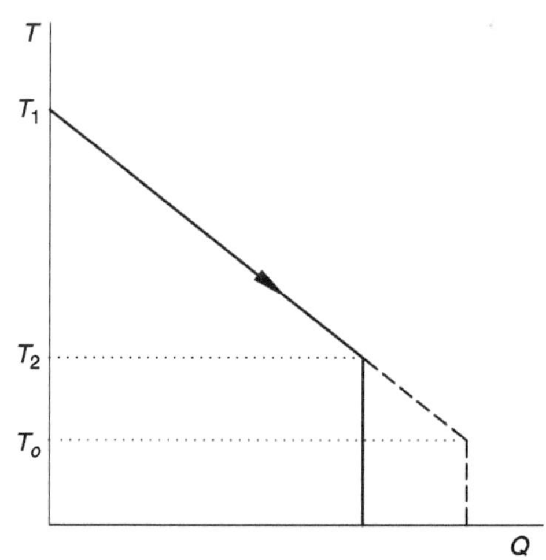

Figure 2.19 Heat transfer as a function of temperature for a single-phase heating medium

slightly simplified since it has been assumed that, in the limit, the heating medium can be cooled to the maximum temperature of the working fluid within the Stirling engine. In a practical case, there would have to be a limiting temperature difference between the two fluids, as already indicated in Fig. 2.17. This is normally referred to as the 'pinch point', as shown in Fig. 2.20(b).

To quantify the loss in efficiency due to the poor matching of the working fluid and heat source temperatures during the heat transfer process (as shown in Fig. 2.20(a)), Fig. 2.21 shows the attainable cycle efficiencies of an ideal Stirling engine and an ideal quadlateral cycle, both using an ideal gas and with the minimum temperature of the heat source limited, in both cases, to $\sqrt{T_1T_0}$. As can be seen, the ideal quadlateral cycle is around 50% more efficient over the given temperature range.

However, based on Eqs. 2.13 and 2.14, it can be shown that the work ratio of the ideal Stirling cycle is approximately 20% greater than that of the ideal quadlateral cycle over this

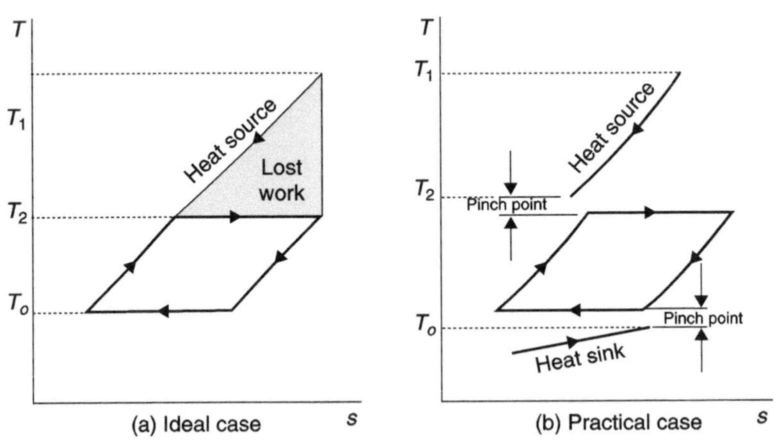

Figure 2.20 Temperature–entropy diagram of ideal Stirling cycle with heat source and sink

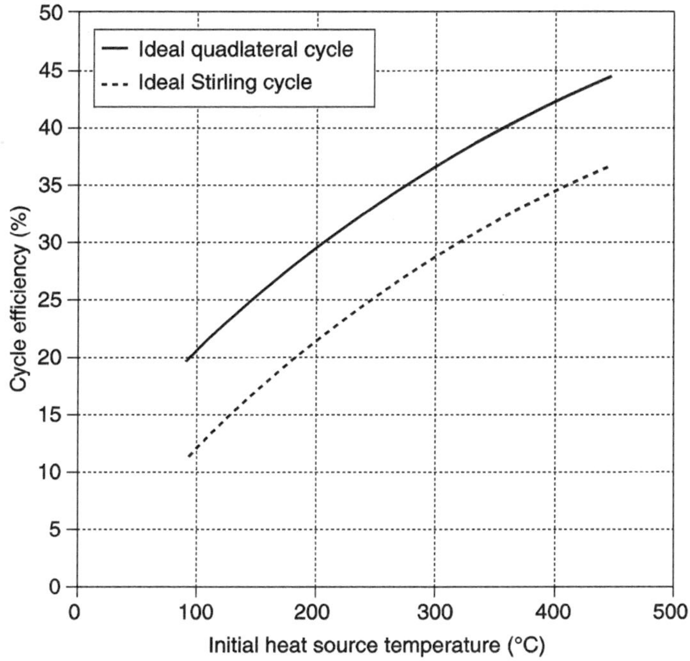

Figure 2.21 **Comparison of performance of ideal quadlateral and Stirling cycles using a perfect gas. Minimum heat source temperature 80 °C, heat rejection temperature 15 °C**

temperature range. Assuming, rather optimistically, expansion and compression efficiencies of 0.9 and 0.8, respectively, in both cycles, and applying Eq. 2.11 to Eqs. 2.4 and 2.14, as also shown in Fig. 2.22, the difference between the practical Stirling cycle and practical quadlateral cycle efficiencies is much reduced over most of the temperature range shown, although neither is capable of useful power output from resource temperatures of less than about 250 °C.

Clearly, good temperature matching between the heating medium and the working fluid is desirable, but poor temperature matching can be compensated by the use of a higher work ratio cycle.

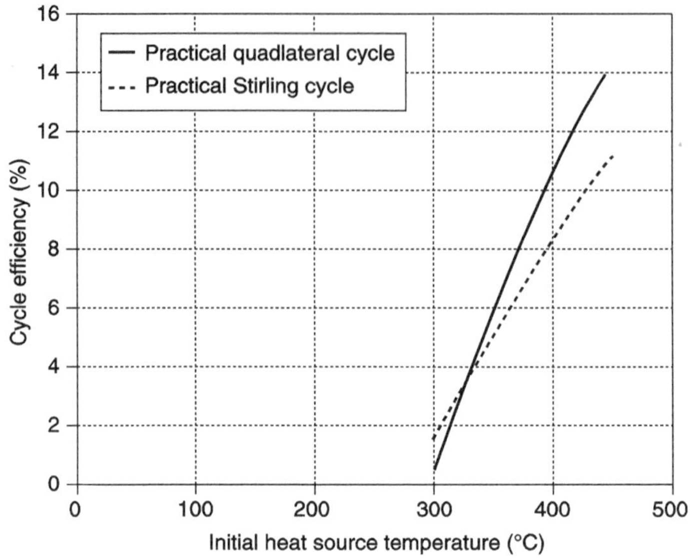

Figure 2.22 **Comparison of performance of practical quadlateral and Stirling cycles. Minimum heat source temperature 80 °C, heat rejection temperature 15 °C**

2.3.3 Higher work ratio cycles

In the preceding sections of this chapter, we have attempted to show that the ideal cycle for power generation from most low-grade heat sources is not that of Carnot, but rather the trilateral or quadlateral concept, depending on the temperature limits of the heat sources. It has also been shown that it is very difficult to even approximate to these when using cycles based on ideal gases as the working fluid, and that the main desiderata for maximising performance are good temperature matching between the heat source and the working fluid, during the heat transfer process between them, and a high cycle work ratio.

It is no coincidence that the first successful heat engines to produce mechanical power were based on the use of water, pressurised by gravity, and then expanded by evaporation to produce steam, because by this means the operating cycle had a work ratio of unity and therefore would produce a positive power output as long as some expansion work was recoverable, regardless of the expander efficiency, which was very low. After further development, this led to the Rankine cycle, with its various improvements, being the dominant means of power generation throughout the nineteenth century and, indeed, in the early years of motor car development, it was not immediately obvious that internal combustion engines would supplant steam engines as the means of motive power. In its most basic form, both the cycle and the system of components, through which it is achieved, are shown in Fig. 2.23.

The main feature of the simple steam Rankine cycle is that because pressurisation of the working fluid is carried out in the liquid phase, where the working fluid is almost incompressible, the feed pump work required is very small. Also, water has the largest ratio of 'latent heat' to 'sensible heat' of any known fluid. This implies that the volume of vapour produced per unit mass is very large compared with that of the liquid entering the boiler. Hence the expansion work derived from expanding the steam created by heating and evaporation is very much greater than that of the pumping work. Accordingly, its work ratio is of the order of 0.999 over the temperature and pressure range of interest in recovering power from lower-temperature heat sources.

The consequences of this are as follows: the real cycle efficiency is reduced from the ideal only by the expander efficiency, which is of the order of 0.8–0.9. Its ideal cycle efficiency is quite close to that of Stirling or Carnot cycles, as shown in Fig. 2.24, although it is significantly less than that of the ideal quadlateral cycle. However, because of its high work ratio, the practical Rankine cycle is far superior

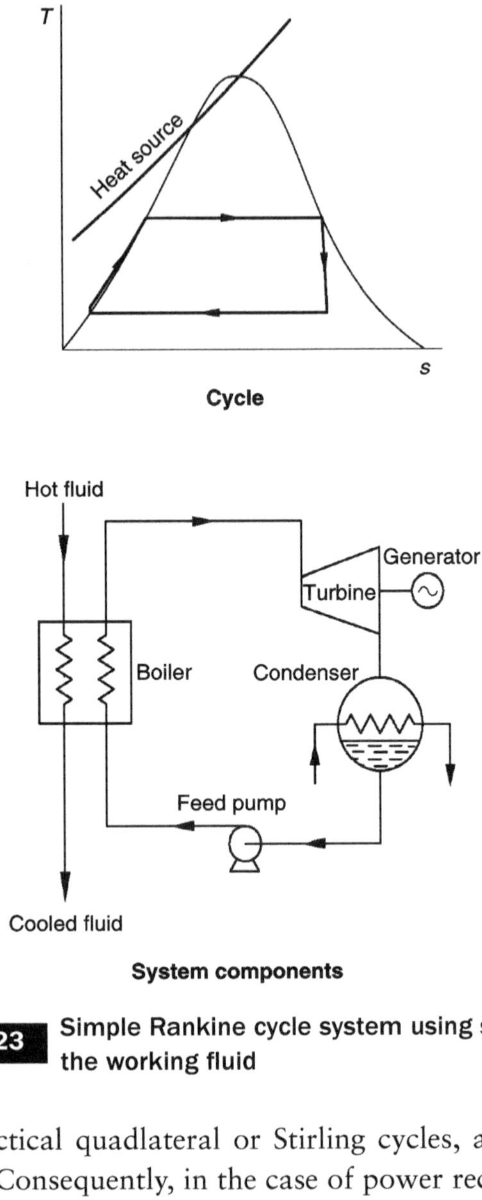

Figure 2.23 Simple Rankine cycle system using steam as the working fluid

to the practical quadlateral or Stirling cycles, as shown in Fig. 2.25. Consequently, in the case of power recovery from low-grade heat, there is little doubt that the Rankine cycle system, and its many subvariants, is the most suitable.

The disadvantage of the Rankine cycle is that the large amount of heat required for evaporation of the water leads

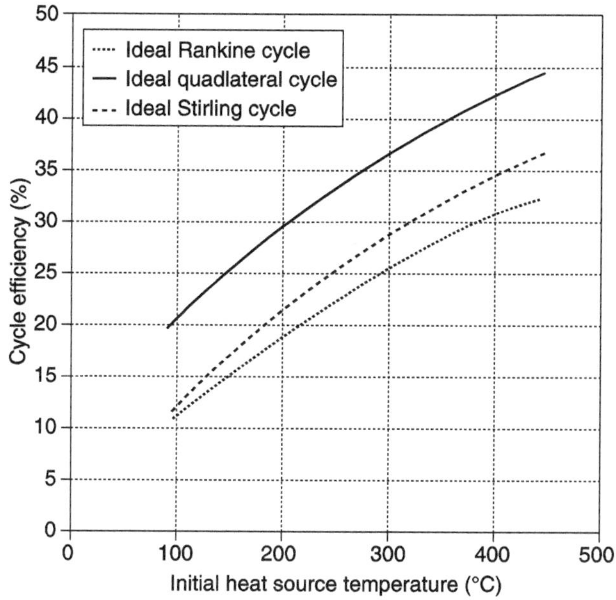

Figure 2.24 Comparison of ideal steam Rankine and ideal gas cycles. Minimum heat source temperature 80 °C, heat rejection temperature 15 °C

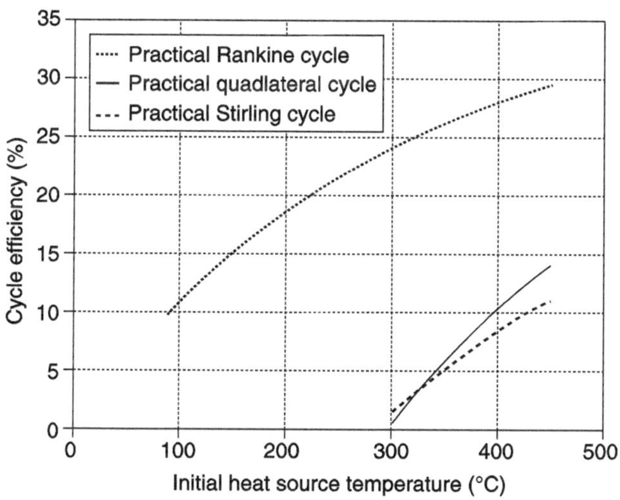

Figure 2.25 Comparison of practical steam Rankine and practical gas cycles. Minimum heat source temperature 80 °C, heat rejection temperature 15 °C

to a poor heat recovery efficiency, and hence the overall conversion efficiency of the system, as shown in Fig. 2.23, is only of the order of 50–60% of the cycle efficiency.

Now consider the system and cycle shown in Fig. 2.26, which we have called a trilateral flash cycle (TFC) system. This has the same number and type of components as a Rankine cycle but differs in that the working fluid is not evaporated in the heater but only heated to its boiling point

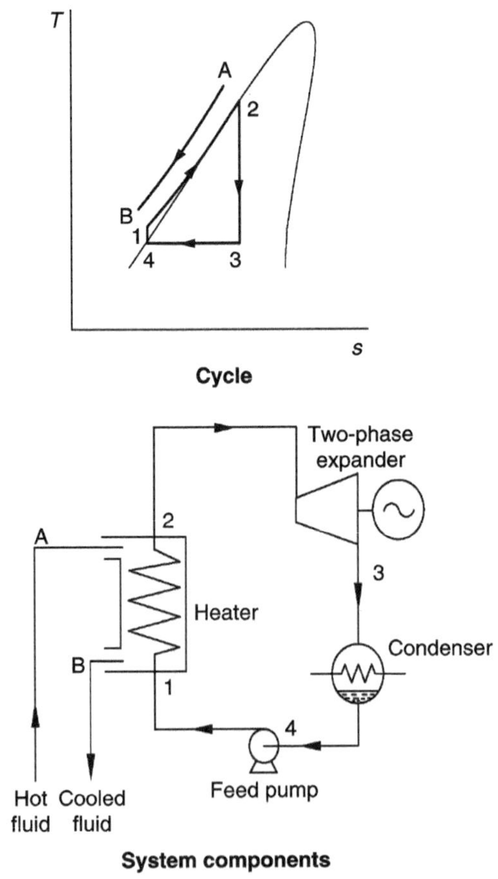

Figure 2.26 Trilateral flash cycle (TFC) system and components

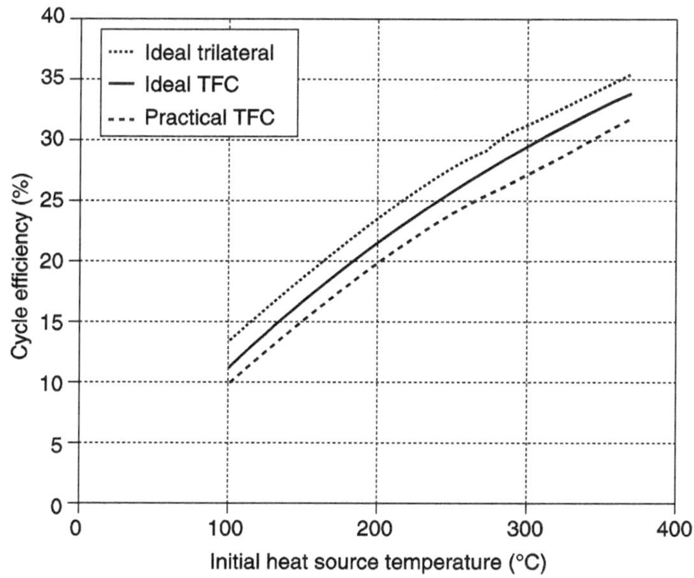

Figure 2.27 **Comparison of performance of trilateral flash cycles (TFC) and ideal trilateral cycle**

and then expanded. As can be seen on the $T-s$ diagram, it appears to be quite close to an ideal trilateral cycle. Its main advantages are that when receiving heat from a single-phase heat source, in the absence of any evaporative heating being needed, it is capable of very high heat recovery efficiencies, while maintaining a high work ratio. Figure 2.27 shows some characteristics of its performance, using water as the working fluid, assuming that there is a 10 °C temperature differential between the working fluid and the heating medium in the fluid heater.

The TFC system, as shown, is clearly the most efficient of those considered, but it has a number of deficiencies, namely:

1. When the heating medium has a limiting lower temperature, such as in the case of geothermal brines containing dissolved solids, from which precipitation must be avoided, IC engine combustion products

containing sulphur and similar contaminants and where there are other uses for the lowest temperature heat, such as for heating purposes. In that case, the temperature matching between the working fluid and the heat source in the heater can be quite poor and the overall conversion efficiency much reduced, as indicated in Fig. 2.28.

2. Two-phase expanders are less efficient than dry vapour expanders.
3. Water, as a working fluid, attains very low pressures at condensing temperatures below 100 °C. This results in problems in maintaining a vacuum in the condenser and very low vapour densities. The latter implies that the volume flows and volume ratios in the expander are very big and result in large and expensive machines and expander design problems, which will be described in a later section.
4. The working fluid heater is very large, both because of the decreasing efficiency of conversion of the heat to power as the heat source temperature decreases and because the mean temperature difference between the working fluid and the heat source is relatively small.
5. The high recovery of heat in the TFC implies a high rejection of heat in the condenser. This implies not only that the condenser will be large but that the coolant flow rates needed to remove the heat will also be large.

The Rankine cycle performance can be improved if account is taken of the fact that not all the heat is transmitted to the working fluid at constant temperature, as is required in the Stirling cycle, to which it was compared in Figs. 2.24 and 2.25. Thus, as indicated in Fig. 2.29, both the cycle and heat recovery efficiencies are increased by reducing the exit temperature of the heating medium to a lower value than

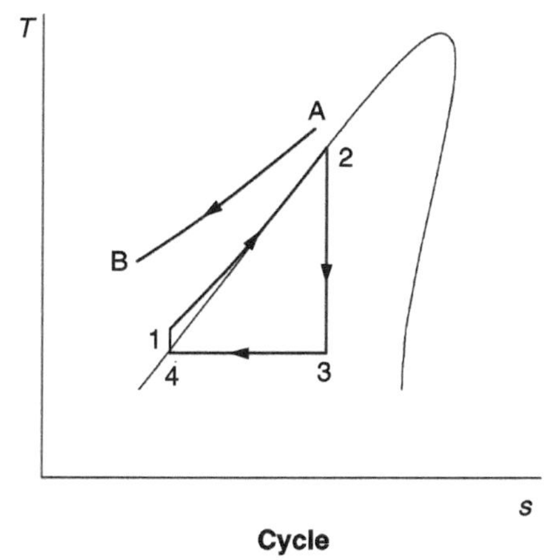

Figure 2.28 TFC temperature matching to a limited minimum temperature heat source

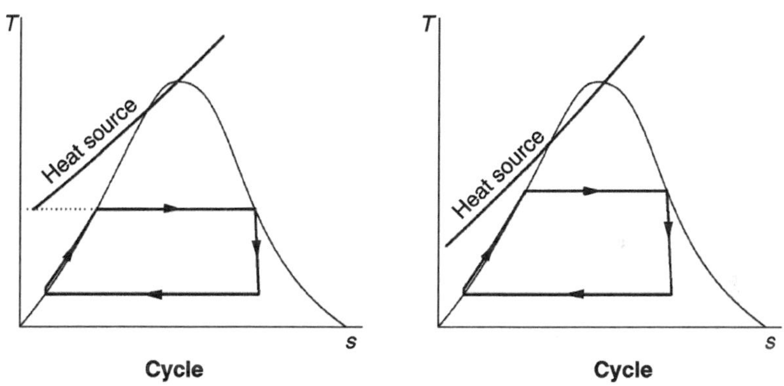

Figure 2.29 Improving the Rankine cycle matching to its heat source

that of the evaporating temperature, thereby obtaining better matching. However, it should be noted that the deficiencies described in item (3), of those enumerated above for a TFC expander, also apply to a Rankine cycle expander.

As will be shown, by using working fluids other than water, many of the deficiencies of both the Rankine and the

TFC systems, as itemised, can be overcome. The differences between them are then reduced to the extent that, in many cases, the Rankine cycle offers a better solution.

2.4 Working fluids other than water

The thermodynamic deficiencies of water as the working medium for vapour power plant have been recognised for longer than most engineers are aware, and it has been reported that Sir Humphrey Davy wrote a note in 1823 suggesting that 'the changes with temperature of the vapour pressure of some substances might enable them to be used as substitutes for steam in an engine', while numerous attempts – some successful – were made to build vapour power plant using a variety of organic working fluids, throughout the nineteenth and early twentieth centuries. However, the individual most responsible for establishing the viability of such systems is, undoubtedly, Lucien Bronicki, who provided the first evidence of a sustained development programme. This has resulted in his being responsible for over 1700 MW of installed Organic Rankine cycle power generating plant to date.

Although liquid metals such as mercury and sodium have been considered, and used, for some applications, especially when recovering power from high-temperature heat sources, for power recovery from low-grade heat, most alternatives to water – proposed or used – are based on hydrocarbons, either of the straight chain type, such as the paraffins (alkanes) and their isomers, or aromatics of the ring type, such as benzene. Halogenated versions of these, such as some of the common refrigerants, are also used, but chlorine-based compounds are now no longer permitted and currently there are only a very limited number of fluorinated alternatives. One of the more recent of these,

R245fa (1,1,1,3,3-pentafluoropropane) is gaining in popularity but concerns over the environmental impact of its manufacture may limit the period for which it will be permitted. Pure hydrocarbons, such as isobutane and isopentane, pentane, hexane and their mixtures and toluene, at higher temperatures, are commonly used in operating systems, despite their greater flammability because they are regarded as more environmentally acceptable. They are also more thermally stable than refrigerants and therefore preferred for higher-temperature systems.

Essentially, all substances exhibit roughly similar characteristics in the temperature and pressure region where they change from liquid to vapour, as shown by the saturation envelope in Fig. 2.30, when plotted on temperature–entropy coordinates. A significant variable is, however, the slope of the saturated vapour line. It has been shown that this is roughly dependent on the number of atoms in the molecule of the substance. Hence water and ammonia, which have

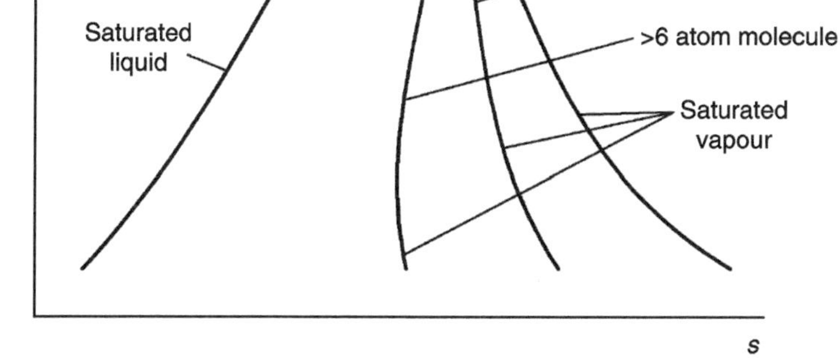

Figure 2.30 Temperature–entropy diagram for various working fluids

3-atom molecules, have a saturated vapour line with a strong negative slope, whereas pentane, C_5H_{12}, which has a 17-atom molecule, has a saturated vapour line with a strong positive slope. Between these, R134a, tetrafluoroethane, a commonly used refrigerant with an 11-atom molecule, has a saturated vapour line with a nearly vertical slope.

It follows that provided a fluid can be found with the right critical temperature, it is possible to match a given heat source to a supercritical, superheated or saturated vapour cycle, as shown in Fig. 2.31. Moreover, depending on how

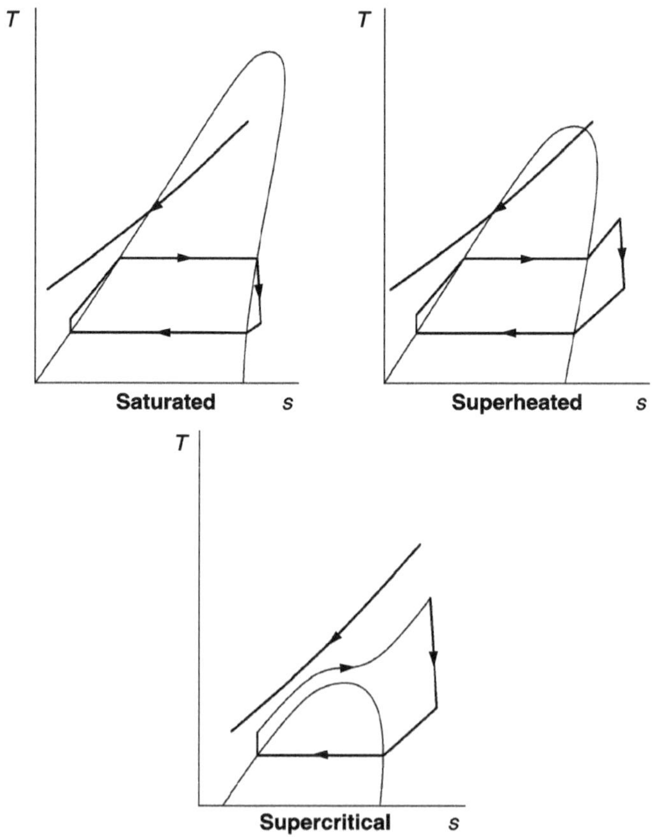

Figure 2.31 Matching the cycle to the heat source with saturated, superheated and supercritical cycles

close the evaporation temperature is to the critical temperature of the working fluid, the proportions of heat transferred in feed heating, evaporation and superheat in a subcritical cycle can be varied to maximise the recoverable power by best matching the cycle to the heat source, as shown in Fig. 2.31.

As can be seen from Figs. 2.30 and 2.31, if the working fluid saturated vapour line has a positive slope on $T-s$ coordinates and expansion begins with saturated or superheated vapour, then the vapour must leave the expander in the superheated condition. This is particularly advantageous for turbines, where the absence of liquid leads to higher efficiencies than is possible with steam, where the vapour normally leaves the turbine as wet vapour. However, it leads to the need for a desuperheater to remove heat at higher temperatures, prior to the start of condensation. One way to do this more efficiently is to use most of the desuperheat to preheat the liquid leaving the feed pump in a recuperative or regenerative heat exchanger, as shown in Fig. 2.32. Recuperation can also be used in supercritical cycles, if required.

The relative merits of the various subvariants of the basic Rankine cycle and how they affect temperature matching between the heat source and the working fluid in the boiler, as well as the work ratio, will be discussed in more detail in Chapter 5.

2.5 Fluid properties

The estimation of fluid thermodynamic and transport properties has been extensively investigated, mainly by chemical engineers and physical chemists employed by the chemical and petroleum industries, where accurate evaluation

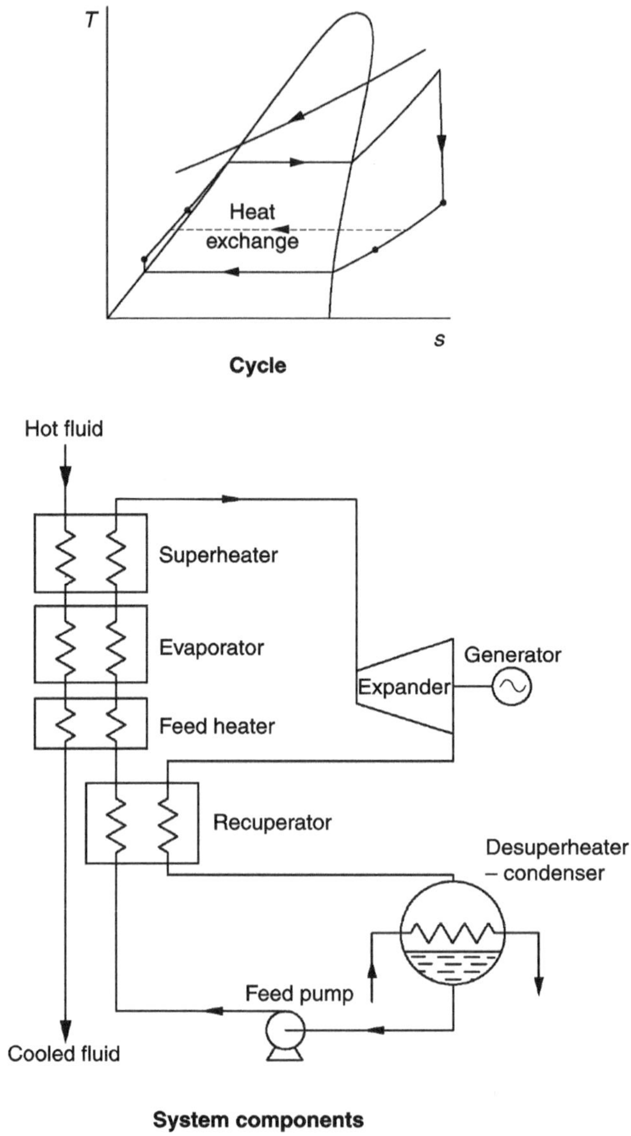

Figure 2.32 Improving the cycle efficiency with a recuperative heat exchanger

of temperatures and pressures and heat supply and rejection rates required for the efficient implementation of chemical processes is essential for optimising the design of chemical plant. Much of this has been summarised by Reid, Prausnitz and their successors in a series of editions and, until relatively recently, would-be investigators and designers of thermal power plant, based on the use of volatile working fluids, needed to develop calculation procedures, based on such publications, in order to carry out power plant cycle analyses and fluid expander and compressor design. A list of the key properties of some commonly used working fluids is given in Appendix 1, while a typical method, formerly used by the authors for mixture property estimation, is given in Appendix 2. Fortunately, today, a number of commercial software packages are available that can readily be used to estimate any fluid properties required without any detailed knowledge of how these are derived. The most widely used of these is the NIST package of subroutines, produced by the US National Institute of Standards. This is readily available, is continually being updated to include more fluids and improved estimation procedures, and includes a database of virtually all known substances currently used or considered as suitable working fluids for low-grade heat power generating plant, both as pure fluids and as mixtures. Despite these calculating procedures now being available, a few guiding principles must be understood in order to guide the plant designer in best fluid choice.

Water, ammonia and other fluids with strong polar forces within their molecules have much higher critical pressures than most organic fluids, for example 221.2 bar for water and 113.33 bar for ammonia. Also, their pressure–volume–temperature relationships do not conform so well with those of organic fluids, in which there are only weak or no polar forces.

In contrast, most organic fluids have critical pressures between 30 and 50 bar and there is a marked similarity between their properties, when these are divided by their critical values to obtain their reduced values. Thus, temperature is expressed as reduced temperature, $T_r = T/T_c$, and pressure as reduced pressure, $p_r = p/p_c$, etc. Their most significant features are as follows:

1. The normal boiling point at atmospheric pressure is roughly constant with a reduced temperature of approximately 0.6, i.e. $T_{boiling\ point}/T_{critical} \approx 0.6$.
2. There is a roughly equal correspondence between reduced vapour pressure and its corresponding reduced saturation temperature for most fluids. The relationship is not linear, since the vapour pressure increases much more rapidly towards its critical value, at reduced temperatures above $T_r = 0.95$. However, there is a near-linear relationship between the logarithm of vapour pressure and the reciprocal of its corresponding absolute temperature, as shown in Fig. 2.33. This is defined by the Cox-Antoine equation: $\ln p = A - (B/T)$, where A and B are constants which are separately defined for each fluid.
3. Since critical pressures do not vary much between different fluids, those with a high critical temperature (or boiling point) will have a low vapour pressure at normal condensing temperatures and vice versa.
4. At the atmospheric boiling point, the entropy of vaporisation per mole is roughly constant for all fluids (Trouton's rule). Since the enthalpy of vaporisation $h_{fg} = T \cdot s_{fg}$, it follows that fluids with a low molecular weight, such as water, have a large enthalpy of vaporisation and, consequently, a high specific enthalpy in the vapour phase. Accordingly, the expansion of low molecular

weight fluid vapours is associated with large drops in specific enthalpy and hence high velocities and vice versa.

A selection of working fluids most commonly used or considered for systems for the recovery of power from low-grade heat is shown in Fig. 2.34. The abscissa of this plot has been normalised to give all the fluids shown a similar slope. Hence the entropy of each fluid does not relate to unit mass. Presentation in this form has been deliberately chosen because it gives the potential power plant designer a better guide to fluid choice, when considering matching the cycle to the heat source.

The working fluids defined by the prefix 'R' were classified as such because their prime function was as working fluids in refrigeration and air conditioning systems. More generally, there is a trend to replace the 'R' by 'HFC' (hydro fluoro carbon). These are all mainly straight chain hydrocarbons

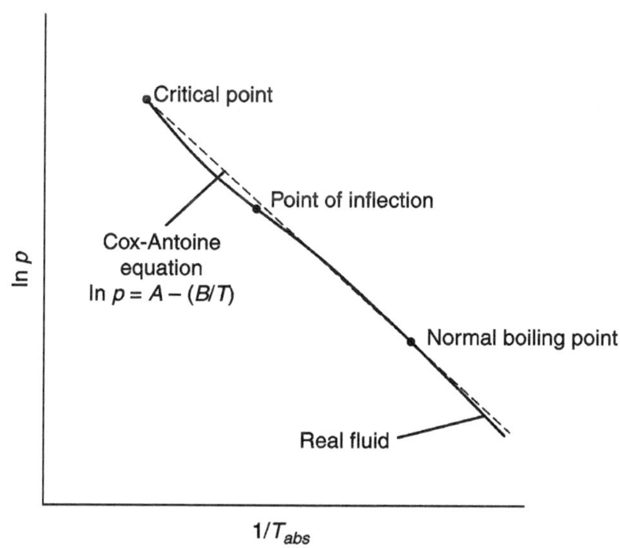

Figure 2.33 The relationship between saturated vapour pressure and temperature for pure fluids

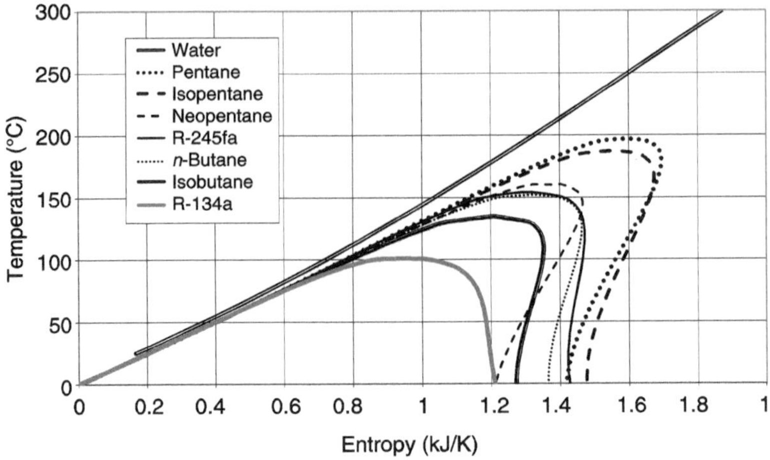

Figure 2.34 **Common working fluids with a saturated liquid line slope approximately equal to that of water**

(alkanes or paraffins), which have the chemical composition C_nH_{2n+2} in which one or more of the hydrogen atoms have been substituted by chlorine, fluorine or, occasionally, bromine. Currently, all chlorine-containing refrigerants are banned except for R124, which is still permitted in North America. The number following defines the chemical composition. Thus the first digit has the value of one less than the number of carbon atoms in the molecule, the second digit one plus the number of hydrogen atoms, and the third digit the number of chlorine, fluorine or bromine atoms, the latter being followed by subscript Br. Thus R124 is chlorotetrafluoroethane and R134a is tetrafluoroethane, the letter 'a' referring to the order in the hydrocarbon molecule in which the fluorine atoms are located, etc.

The selection of these fluids depends on a number of factors. These include the absolute temperature of the heat source, its permitted minimum temperature and its cost, the working fluid critical temperature and the pressure range over which it operates, and the temperature and the nature of the coolant available for removing heat from the condenser. Some aspects of this are best appreciated by the following comparison between R134a and pentane.

Because of its low critical temperature (101.06 °C) and relatively high critical pressure (40.59 bar), an R134a system will operate at high condensing pressures and with large pressure differences between the boiler and condenser. It is best suited to heat sources of relatively low temperature, where good temperature matching can be obtained between the cooling heat source and the working fluid, since a high percentage of the heat transferred will be in the liquid phase. Because of its relatively high relative molecular mass (102.03), the specific enthalpy drop in expansion is comparatively small and hence by default the feed pump work is relatively large, thus robbing the system of some of the thermodynamic advantage gained by good temperature matching in the boiler. The relatively high condensing pressure, however, implies that the vapour is dense and this means that the expander and heat exchangers will be relatively compact and cheap.

Pentane has a relatively high critical temperature (196.55 °C) but a lower critical pressure (33.7 bar), while its relative molecular mass is 72.15. Accordingly, it is best suited to higher-temperature cycles, where it will match better with a higher-temperature heat source. Under these conditions, the feed pump work will be relatively much less and hence its work ratio will be higher than in a R134a system. However, the condensing pressure will be relatively low because of its high critical temperature and therefore the vapour density will be low, resulting in larger and more expensive components for the same heat rejection rate as in a R134a system. On the other hand, because a pentane system will best operate at high temperatures, its cycle efficiency will be much higher than that of a R134a system and therefore the power generated per unit of heat transferred will be significantly larger. Hence despite the large size per unit of heat transferred, the cost per unit power output may be less.

Several investigators have proposed screening methods for best fluid selection. However, in our opinion, there are too many factors involved to make such methods foolproof. Detailed analytical studies, which can now be carried out relatively easily with good software, though more time-consuming, are more likely to produce the most reliable results.

More details of some of the most commonly used working fluids are given in Appendix 1.

3

Geometry and manufacture of screw expander rotors

DOI: 10.1533/9781782421900.59

Abstract: This chapter demonstrates a method for screw rotor profile generation that simplifies and improves design procedures. An example is given of its use in the development of a new 'N' rotor profile, which is shown to be superior to other well-known types. Twin-screw expander rotors are effectively helical gears. When these are formed from a hobbing cutter, the hobbing tool and the rotor together constitute a pair of crossed helical gears. The envelope gearing method is used to derive a meshing condition for crossed helical gears, which is then used to create the profile of a hobbing tool. A reverse transformation enables calculation of the rotor profile thus manufactured. Simplification of the main gearing condition leads to a meshing expression for helical gears that may be used for the design of screw expander rotors.

Key words: helical screw expander geometry, screw expander rotor profiles, meshing condition, working volume.

Published by Woodhead Publishing Limited, 2014

3.1 Introduction

To perform effectively, screw expander rotors must meet the meshing requirements of gears, while maintaining a seal along their length to minimize leakage at any position on the band of contact between them. The expander efficiency depends on both the rotor profile and the clearances between the rotors and between the rotors and the expander housing.

A typical screw rotor profile, with a 5–6 lobe configuration for the main and gate rotors, is shown in Fig. 3.1. The sealing line of the meshing rotors, as seen projected in their axial plane, is drawn within the left-hand rotor, while the same line, as seen projected in their transverse plane, is shown in the right-hand rotor. The clearance distribution, between the two rotors, as they would appear on equivalent meshing gears of infinite radius (racks), from which the rotor profiles are generated, is shown in the transverse cross-section, scaled by a factor of 50.

The volume of fluid that can be admitted into a screw expander is defined by its rotor profile, and the inlet port shape and size. A gearing algorithm used for the rotor profiling defines the meshing action. A variety of rotor primary curves can be defined either analytically or by discrete points, which meet the meshing criteria.

A screw expander's efficiency is determined by several key parameters, namely its rotor profile, diameter and length-to-diameter ratio, the internal clearances, and its speed. To obtain the best performance for any application, a designer must be able to predict how this is affected by changing any of these parameters.

Rotors are usually manufactured in specialist machines by the use of formed milling or grinding tools. The machining accuracy achievable today is high, and tolerances in rotor

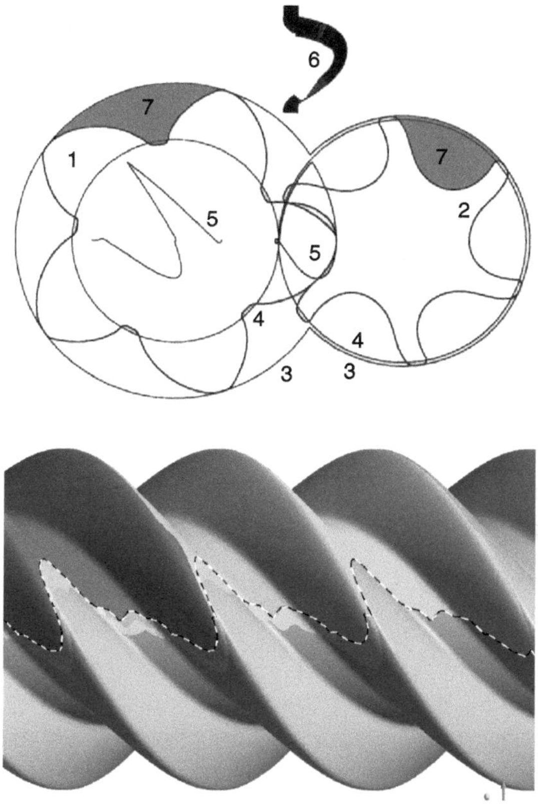

Figure 3.1 Screw rotor profile: 1, main; 2, gate; 3, rotor external and 4, pitch circles; 5, sealing line; 6, clearance distribution; 7, rotor flow area between the rotors and housing

manufacture are of the order of 5 μm around the rotor lobes. Even higher accuracy has been achieved with new vitrifying thread-grinding machine tools, with tolerances within ±3 μm even in large batch production. Thus, from the viewpoint of rotor production only, clearances between the rotors can be as little as 12 μm.

Tight clearances of this order of magnitude result in small internal expander leakage rates. Consequently, further performance improvements are only possible by introducing more refined design principles, to improve the rotor profiles, which maximise the internal flow area through the expander while minimising both the length of the leakage path and the internal friction losses due to relative motion between the contacting rotor surfaces. Also, as more precise manufacture permits rotor clearances to be reduced, the likelihood of direct rotor contact is increased. Friction, resulting from hard rotor contact, leads to local temperature rise and deformation of the gate rotor, thus increasing the contact forces, and ultimately leading to rotor seizure. Hence the profile should be designed to minimise this risk.

3.2 Review of contemporary rotor profiles

Until recently, the design procedures to generate rotor profiles have been mostly to create a primary profile curve on one of the real screw rotors and to generate a corresponding secondary profile curve on the other using a suitable conjugate criterion to ensure that they mesh. Any curve can be used as the primary one, but traditionally a circle is most often used.

The original symmetric circular profile consists of circles only. Lysholm's asymmetric profile introduced a set of cycloids on the high-pressure side, in addition to circles located at the rotor pitch circle, thus forming the first asymmetric screw rotor profile. The SRM (Svenska Rotor Maskiner) asymmetric profile employs an eccentric circle on the low-pressure side of the gate rotor, followed by the SKBK profile introducing the same on the main rotor. In both cases, the curves evolved were given analytically as epicycloids or hypocycloids. The SRM 'D' profile consists exclusively of circles, all of them eccentrically positioned on the main or gate rotor.

Many patented designs followed, specifying primary curves on one rotor, and secondary, generated curves on the other, all probably based on derivations of classical gear theory or similar concepts. More recently, circles have been gradually replaced by other curves, such as ellipses in the Fusheng profiles, parabolae in the Compair and Hitachi profiles, and hyperbolae in the Hanbel hyper profiles. The hyperbola in the latest profile seems to provide the best ratio of rotor displacement to seal line length.

A more recent method to generate screw rotor profile curves is to use imaginary or non-physical rotors. Since all gearing equations are independent of the coordinate system in which they are expressed, it is possible to define primary arcs as given curves using a coordinate system, which is independent of both rotors. By this means, in many cases, the defining equations may be simplified. Also, the use of one coordinate system to define all the curves further simplifies the design process. Typically, a template is specified in a rotor independent coordinate system. This is valid for a rotor of infinite radius, which is a rack. From this, a secondary arc for the other rotor is obtained by what is known as 'rack generation'.

As shown in Fig. 3.1, an efficient screw expander needs a rotor profile, which has a large flow cross-sectional area, a short sealing line, and a small blow-hole area. The larger the cross-sectional area, the higher the flow rate for the same rotor sizes and rotor speeds. Shorter sealing lines and a smaller blow-hole reduce leakages. Higher flow and smaller leakage rates both improve the expander volumetric efficiency, which is the rate of flow delivered as a fraction of the sum of the flow plus leakages. This in turn increases the adiabatic efficiency, because less power is lost due to gas bypassing the working chamber.

The optimum choice of blow-hole and flow areas depends on the expander duty. Thus, for low-pressure differences, the leakage rate will be relatively small, and hence the gains achieved by a large cross-sectional area may outweigh the losses associated with a larger blow-hole. Similar considerations determine the best choice for the number of lobes, since fewer lobes imply greater flow area but increased pressure differences between them.

The search for new profiles has been both stimulated and facilitated by recent advances in mathematical modelling and computer simulation. These analytical methods may be combined to form a powerful tool for process analysis and optimisation, thereby eliminating the earlier approach of intuitive changes, verified by tedious trial-and-error testing. As a result, this approach to the optimum design of screw rotor lobe profiles has evolved substantially over the past few years, and is likely to lead to further improvements in expander performance in the near future.

Traditionally, rotor pairs were manufactured with a 4/6 configuration, i.e. four lobes on the main rotor and six on the gate rotor, all with the same outer diameter. This configuration is a compromise, which has favourable features for both dry and oil-flooded expander applications. However, other configurations, such as 5/6 and 5/7, and recently 4/5 and 3/5, are becoming increasingly popular. Five lobes in the main rotor are suitable for higher expander pressure ratios, especially if combined with larger helix angles. The 4/5 arrangement is the best combination for moderate pressure ratios. The 3/5 is favoured for low-pressure ratio, high-speed applications, because it offers a high gear ratio between the gate and main rotors, thus reducing the drive shaft speed required.

Figure 3.2 shows pairs of screw expander rotors plotted together for comparison. They are given by their commercial names or by names that denote their patent.

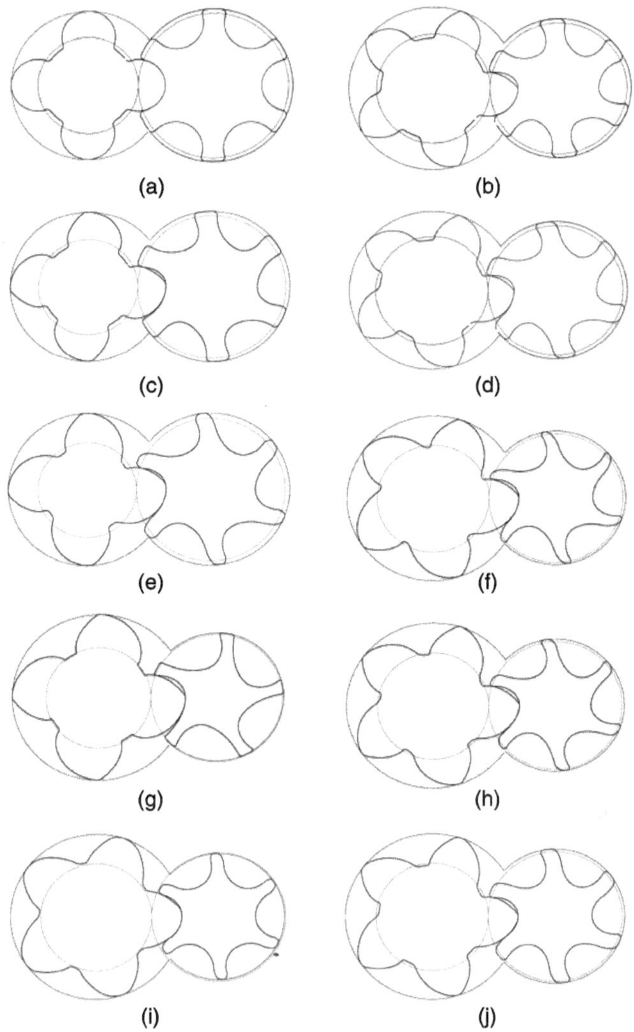

Figure 3.2 **Most popular screw compressor rotors. (a) Symmetric circular profile (Rooney, 1952); (b) Lysholm's asymmetric profile (Lysholm, 1967); (c) SRM 'A' profile (Shibbye, 1979); (d) SKBK profile (Amosov *et al.*, 1977); (e) SRM 'D' profile (Astberg, 1982); (f) Fusheng profile (Lee, 1988); (g) Compair profile (Hough *et al.*, 1984); (h) Hyper profile (Chen, 1995); (i) Rinder's profile (Rinder, 1987); (j) 'N' profile (Stosic, 1996)**

3.3 Screw expander rotor geometry

The rotor lobe profiles have to be defined together with the remaining rotor parameters before the rotor and housing geometry can be fully specified. To explain the rotor profiling principles, the so-called 'demonstrator' defines a default profile. This can include any realistic combination of numbers of lobes in the main and gate screw rotors. Moreover, any known or even a completely new profile can be generated from it, with little modification. Such profiles must, of course, satisfy geometric constraints in order to obtain a realistic shape.

3.3.1 *The envelope method as a basis for profiling screw rotors*

The envelope method is used here as a basis for generating a screw expander rotor profile. The method states that two surfaces will mesh if each generates or envelops the other under a specified relative motion. This is becoming increasingly popular. Although the generation of screw expander rotors can be regarded as a two-dimensional problem, a three-dimensional approach is given here as a general starting point. This also provides an opportunity to develop rotor tools from the same equations.

Screw expander rotors are treated here as crossed helical gears with non-parallel and non-intersecting axes, as shown in Fig. 3.3. x_{01}, y_{01} and x_{02}, y_{02} are the point coordinates in cross-section, using the coordinate system shown for the main and gate rotors. Σ is the angle between the rotational axes z_1 and z_2. Rotation of either rotor shaft around its z axis describes the natural rotor movement in its bearings. While the main rotor rotates through angle θ, the gate rotor rotates through angle $\tau=(r_{1w}/r_{2w})\theta=(n_2/n_1)\theta$, where r_w and n are the

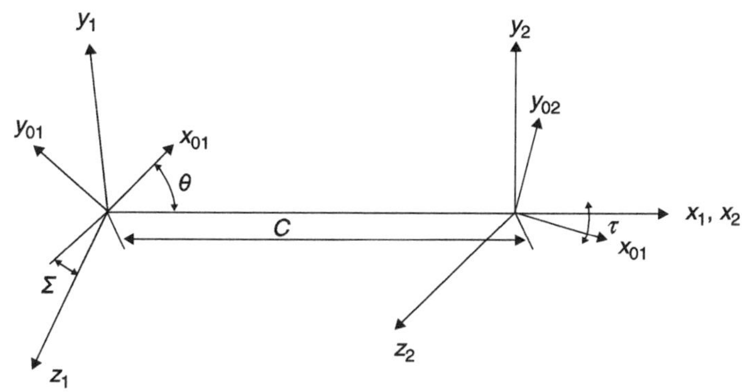

Figure 3.3 **Coordinate system of helical gears with non-parallel and non-intersecting axes**

pitch circle radii and number of rotor lobes respectively, with subscripts '1' and '2' defining the main and gate rotors, while 'w' refers to the pitch circle.

The procedure starts with a given, or generating, surface $r_1(t, \theta)$ for which the meshing, or generated, surface is to be determined:

$$r_1 = r_1(t, \theta) = [x_1, y_1, z_1]$$
$$= [x_{01} \cos \theta - y_{01} \sin \theta, x_{01} \sin \theta + y_{01} \cos\theta, p_1\theta] \qquad [3.1]$$

$$\frac{\partial r_1}{\partial t} = \left[\frac{\partial x_1}{\partial t}, \frac{\partial y_1}{\partial t}, 0\right]$$
$$= \left[\frac{\partial x_{01}}{\partial t}\cos\theta - \frac{\partial y_{01}}{\partial t}\sin\theta, \frac{\partial x_{01}}{\partial t}\sin\theta + \frac{\partial y_{01}}{\partial t}\cos\theta, 0\right] \qquad [3.2]$$

$$\frac{\partial r_1}{\partial \theta} = \left[\frac{\partial x_1}{\partial \theta}, \frac{\partial y_1}{\partial \theta}, p_1\right] = \left[-y_{01}, x_{01}, p_1\right] \qquad [3.3]$$

where p is the rotor lead given for unit rotor rotation angle, and t is the rotor parameter. A family of such generated surfaces is given in parametric form by $r_2(t, \theta, \tau)$, where t is the profile parameter and θ and τ are motion parameters:

$$
\begin{aligned}
r_2 = r_2(t, \theta, \tau) &= [x_2, y_2, z_2] \\
&= [x_1 - C, y_1 \cos \Sigma - z_1, \sin \Sigma, y_1, \sin \Sigma + z_1 \cos \Sigma] \\
&= [x_{02} \cos \tau - y_{02} \sin \tau, x_{02} \sin \tau + y_{02} \cos \tau, p_2 \tau]
\end{aligned} \quad [3.4]
$$

$$
\begin{aligned}
\frac{\partial r_2}{\partial \tau} &= [-y_2, x_2, p_2] = [x_{02} \sin \tau + y_{02} \cos \tau, x_{02} \cos \tau - y_{02} \sin \tau, p_2] \\
&= [p_1 \theta \sin \Sigma - y_1 \cos \Sigma, p_2 \sin \Sigma + (x_1 - C) \cos \Sigma, p_2 \cos \Sigma \\
&\quad - (x_1 - C) \sin \Sigma]
\end{aligned} \quad [3.5]
$$

where C is the rotor centre distance.

The envelope equation, which determines the meshing between the surfaces r_1 and r_2 is defined as:

$$
\left(\frac{\partial r_2}{\partial t} \times \frac{\partial r_2}{\partial \theta} \right) \cdot \frac{\partial r_2}{\partial \tau} = 0 \quad [3.6]
$$

Together with Eqs. 3.1 and 3.4 for these surfaces, this completes a system of equations. If a generating surface, 1, is defined by the parameter t, the envelope may be used to calculate another parameter, θ, as a meshing condition. This is now a function of t and defines a generated surface, 2, now the function of both t and θ. The cross product in the envelope equation represents the normal to the surfaces, and $\partial r_2 / \partial \tau$ is the relative sliding velocity between surfaces 1 and 2. Taken together, these define the common tangential point of contact of these two surfaces. Since it is a requirement of the given coordinate system that the value of the scalar triple product must equal zero, and since the relative velocity may be represented in both coordinate systems concurrently, the meshing condition can conveniently be defined as:

$$
\left(\frac{\partial r_1}{\partial t} \times \frac{\partial r_1}{\partial \theta} \right) \cdot \frac{\partial r_1}{\partial \tau} = -\left(\frac{\partial r_1}{\partial t} \times \frac{\partial r_1}{\partial \theta} \right) \cdot \frac{\partial r_2}{\partial \tau} = 0 \quad [3.7]
$$

Insertion of the previous expressions into this envelope condition leads to:

$$\begin{aligned}&\left[C-x_1+(p_1-p_2)\cot\Sigma\right]\left(x_1\frac{\partial x_1}{\partial t}+y_1\frac{\partial y_1}{\partial t}\right)\\&+p_1\left[p_1\theta\frac{\partial y_1}{\partial t}+(p_2-C\cot\Sigma)\frac{\partial x_1}{\partial t}\right]=0\end{aligned}\qquad [3.8]$$

This equation is applied here to derive the meshing condition for crossed helical gears of uniform lead with non-parallel and non-intersecting axes. The method constitutes a gear generation procedure that can be applied generally. It can be used for the design of screw expander rotors, which are effectively helical gears with parallel axes. Formed tools for rotor manufacturing are crossed helical gears on non-parallel and non-intersecting axes with a uniform lead, as in the case of hobbing, or with no lead, as in formed milling and grinding. Templates for rotor inspection are the same as planar rotor hobs. In all these cases the tool axes do not intersect the rotor axes. Screw rotor gearing can then be shown as an elementary example of its use, while a procedure for generating a hobbing tool is a complex case.

The shaft angle Σ, centre distance C, and the unit leads of the two crossed helical gears p_1 and p_2 are not interdependent. The meshing of crossed helical gears is still preserved because both gear racks have the same normal cross-sectional profile, and the rack helix angles are related to the shaft angle as $\Sigma=\psi_{rw1}+\psi_{rw2}$. This is achieved by an implicit shift of the gear racks in the x direction, thereby forcing them to adjust to the appropriate rack helix angles. This includes special cases, such as that of gears that may be orientated so that the shaft angle is equal to the sum of the gear helix angles (i.e. $\Sigma=\psi_1+\psi_2$). Furthermore, the centre distance is then equal to the sum of the gear pitch radii (i.e. $C=r_{1w}+r_{2w}$).

Pairs of crossed helical gears may have either both helix angles of the same sign or each of opposite sign, left- or right-handed, depending on the their lead and shaft angles.

Equation 3.8 can only be solved by numerical methods. For the given parameter t, the coordinates x_{01} and y_{01}, and their derivatives $\partial x_{01}/\partial t$ and $\partial y_{01}/\partial t$ are known. A guessed value of parameter θ is then used to calculate x_1, y_1, $\partial x_1/\partial t$ and $\partial y_1/\partial t$. A revised value of θ is then derived, and the procedure repeated until the difference between two consecutive values becomes sufficiently small.

For given transverse coordinates and derivatives of gear 1 profile, θ can be used to calculate the x_1, y_1 and z_1 coordinates of its helicoid surface. The gear 2 helicoid surface can then be calculated. Coordinate z_2 can then be used to calculate τ, and finally, its transverse profile point coordinates x_2 and y_2 can be obtained.

A number of cases can be identified from this analysis:

1. When $\Sigma = 0$, the equation meets the meshing condition of screw expander rotors and also helical gears with parallel axes. For such a case, the gear helix angles have the same value, but opposite sign and the gear ratio $i = p_2/p_1$ is negative. The same equation may also be applied for the generation of a rack formed from those gears. In addition, it describes a formed planar hob, front milling tool and a template control instrument.
2. If a disc-formed milling or grinding tool is considered, it is sufficient to set $p_2 = 0$. This is a singular case when free rotation of the tool does not affect the meshing process. Therefore, rotor profile generation by reverse transformation cannot be obtained directly.
3. The full scope of the meshing condition is required for the generation of the profile of a formed hobbing tool. Therefore, the most complicated type of gear can be generated from Eq. 3.8.

3.3.2 Direct digital simulation

A standard intrinsic feature of perhaps every existing design software package is to extract the overlapping areas of planar shapes or volumes of spatial bodies. This can be conveniently used to find a corresponding meshing profile from a given profile. This subtraction may either be static, when the overlapping elements are extracted from the stationary assembly, or dynamic, when the subtraction is performed for a sequence of succeeding positions to give a generated profile or body over a complete cycle. All kinds of relative motion can be imposed between the elements, and thus a variety of gear pairs can be designed for parallel, intersecting or non-intersecting and non-parallel axes. A known profile will therefore cut the required shape from a blank with an arbitrary starting shape. This feature enables non-specialists to generate their own rotor profiles without knowledge of comprehensive mathematical methods or use of specialized generation software tools. The application of direct digital simulation to generate a gate rotor from its main counterpart is presented in Fig. 3.4.

3.3.3 Meshing of rotor profiles

As already stated, screw expander rotors have parallel axes and a uniform lead, and they are therefore a form of helical gears (Fig. 3.5). The centre distance for this particular case is $C = r_{1w} + r_{2w}$. Rotors make line contact, and the meshing criterion in the transverse plane perpendicular to their axes is the same as that of spur gears. Although spur gear meshing fully defines helical screw rotors, it may be more convenient to use the envelope condition for crossed helical gears and simplify by setting $\Sigma = 0$, to get the required meshing condition.

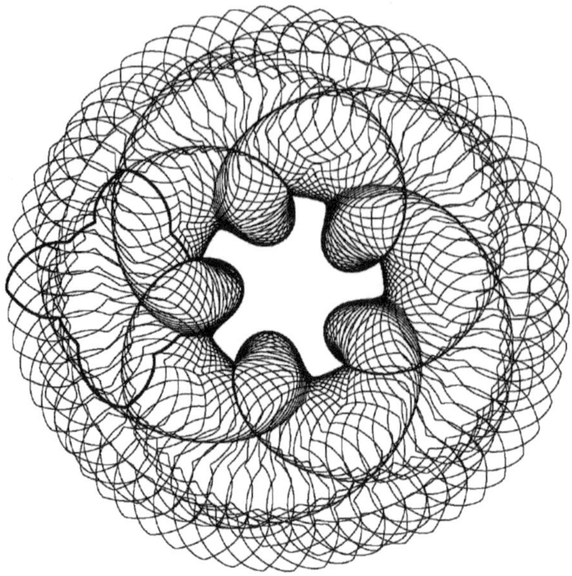

Figure 3.4 Example of a gate rotor enveloped by its main counterpart using direct digital simulation

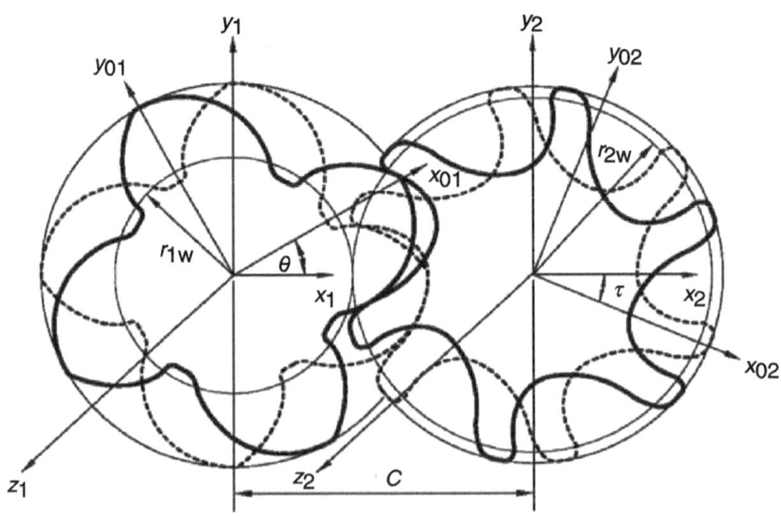

Figure 3.5 Screw expander rotors with parallel shafts and their coordinate systems

To start the procedure of rotor profiling, the profile point coordinates in the transverse plane of one rotor, x_{01} and y_{01}, and their first derivatives, either $\partial x_{01}/\partial t$ and $\partial y_{01}/\partial t$ or $\partial y_{01}/\partial x_{01}$, must be known. This profile may be specified for either the main or gate rotors or, sequentially, for both. The primary profile may also be defined as a rack.

Since $\Sigma=0$, the general meshing condition for a screw expander rotor presented in Eq. 3.8 reduces to:

$$\frac{\partial y_{01}}{\partial x_{01}}\left(ky_{01}-\frac{C}{i}\sin\theta\right)+kx_{01}+\frac{C}{i}\cos\theta=0 \qquad [3.9]$$

where $i=p_2/p_1$ and $k=1-1/i$. Equation 3.9 can only be solved numerically. Once obtained, the distribution of θ along the profile may be used to calculate the meshing rotor profile point coordinates, as well as to determine the sealing lines and paths of proximity between the two rotors. Rotor rack coordinates may also be calculated from the same θ distribution.

Since $\tau=\theta/i$ for parallel axes, the meshing profile equations of the gate rotor in the transverse plane are calculated as:

$$\begin{cases} x_{02}=x_{01}\cos k\theta-y_{01}\sin k\theta-C\cos\dfrac{\theta}{i} \\ y_{02}=x_{01}\sin k\theta+y_{01}\cos k\theta+C\sin\dfrac{\theta}{i}. \end{cases} \qquad [3.10]$$

Rack coordinates can be obtained uniquely if the rack-to-rotor gear ratio i tends to infinity:

$$\begin{cases} x_{0r}=x_{01}\cos\theta-y_{01}\sin\theta, \\ y_{0r}=x_{01}\sin\theta+y_{01}\cos\theta-r_1\theta \end{cases} \qquad [3.11]$$

Conversely, if the gate rotor curves are given, generated curves will be placed on the gate rotor, and similar equations with substituted indices will be used to generate the main rotor profile.

However, if the primary curves are given on the rack, their coordinates x_{0r} and y_{0r}, as well as their first derivatives, $\partial x_{0r}/\partial t$ and $\partial y_{0r}/\partial t$ or $\partial y_{0r}/\partial x_{0r}$, must be known and the generated curves will be calculated at the rotors as:

$$\begin{cases} x_{01} = x_{0r}\cos\theta - (y_{0r} - r_{1w})\sin\theta \\ y_{01} = x_{0r}\sin\theta + (y_{0r} - r_{1w})\cos\theta \end{cases} \qquad [3.12]$$

after the meshing condition is obtained from

$$\frac{dy_{0r}}{dx_{0r}}(r_{1w}\theta - y_{0r}) - (r_{1w} - x_{0r}) = 0 \qquad [3.13]$$

The rack meshing condition θ can be solved directly and does not require a numerical procedure for its evaluation. This is another advantage of the rack generation method.

The method shown to solve the meshing condition either numerically or directly enables a variety of primary arc curves to be introduced and basically offers a general procedure. Moreover, numerical derivation of the primary arcs enables a general approach when only the coordinates of the primary curves need be known and not their derivatives. In such a case, any analytical functions and even discrete point functions can be used as primary arcs. This approach further simplifies the procedure.

The sealing line of screw expander rotors is somewhat similar to the gear contact line. Since there exists a clearance gap between the rotors, the sealing line is a line consisting of points corresponding to the most proximate rotor position. Its coordinates are x_1, y_1 and z_1, and they are calculated for the same θ distribution. A convenient way to obtain an interlobe clearance gap is to consider it as the shortest distance between the two rotor racks of the main and gate rotor sealing points in the cross-section normal to the rotor helicoids.

The rotor racks, obtained from the rotors by the reverse procedure, can include all manufacturing and positioning imperfections. Therefore, the resulting clearance distribution may represent real life expander clearances. From normal clearances, a transverse clearance gap may be obtained by appropriate transformation.

As has already been mentioned in this section, the profile point coordinates in the transverse plane of one rotor, x_{01} and y_{01}, and their first derivatives, $\partial x_{01}/\partial t$ and $\partial y_{01}/\partial t$ or $\partial y_{01}/\partial x_{01}$, must be known. Since the rotor coordinates are integrals of their derivatives, it appears that only the derivatives must be known to be able to generate the whole rotor profile from their initial condition.

For further analysis of the expander geometry, several generic definitions are introduced here. The centre distance between the rotor axes is $C = r_{1w} + r_{2w}$, where r_{1w} and r_{2w} are the radii of the main and gate rotor pitch circles, respectively. The rotor gear ratio is $i = r_{2w}/r_{1w} = n_2/n_1$, where n_1 and n_2 are now the number of lobes on the main and gate rotors. Since the screw expander rotors are three-dimensional bodies, the helix angle ψ is defined at some rotor radius, while ψ_w corresponds to the pitch circle. Then $\tan\psi/\tan\psi_w = r/r_w$. The helix angle defines the rotor lead h, which can be given relative to the unit angle from $p = h/2\pi$. The rotor length L, wrap angle φ and lead are interrelated by $L/\varphi = h/2\pi = p$. If the rotors are unwrapped, a simple relation between the wrap and helix angles can be established from $\tan\psi_w = \varphi r_w/L = 2\pi r_w/h$.

The rotor displacement is the product of the rotor length and its cross-sectional area, which is denoted by the number 7 in Fig. 3.1.

3.4 Features of 'N' rotor profiles

The following is a detailed presentation of rotor creation by both rotor and rack generation procedures. The rotor profile in the first case is a hypothetical one, and has not yet been used in expander practice. However, this profile has been used frequently for the purpose of education. Furthermore, this profile may conveniently be used as a basis for individual development of screw expander rotors. The second example is an 'N' rotor generated from its rack and this profile is protected by international patents.

3.4.1 Demonstrator rotor profile

The primary or generating lobe profile of the demonstrator consists only of circles. It is given on the main and gate rotors, and the profile is divided into several segments (Fig. 3.6). The lobe segments of this profile are essentially parts of circles on one of the rotors and trochoid curves corresponding to the circles on the opposite rotor.

The following summarizes the specific expressions for the x,y coordinates of the lobe profiles for the main screw rotor, relative to the centre of the rotor θ_1. The pitch radii, r_{1w} and r_{2w}, are given together with the rotor radii r_0, r_1, r_2, r_3 and r_4. The external and internal radii are calculated as $r_{1e}=r_{1w}+r$ and $r_{1i}=r_{1w}-r_0$ for the main and gate rotors, respectively (Fig. 3.6).

When all segments of the main rotor are known, they are used as source curves. The gate rotor lobe is now generated by the meshing procedure described in the previous section.

Although simple and almost generic, the demonstrator profile contains all the features that characterize modern screw rotor profiles. The pressure angles on both the flat and round profile lobes are not zero, which is a prerequisite for successful manufacturing. The profile is generated by curves and not by points, which enhance its manufacturability even further. By changing its parameters, r_1, r_2, r_3 and r_4 (Fig. 3.6),

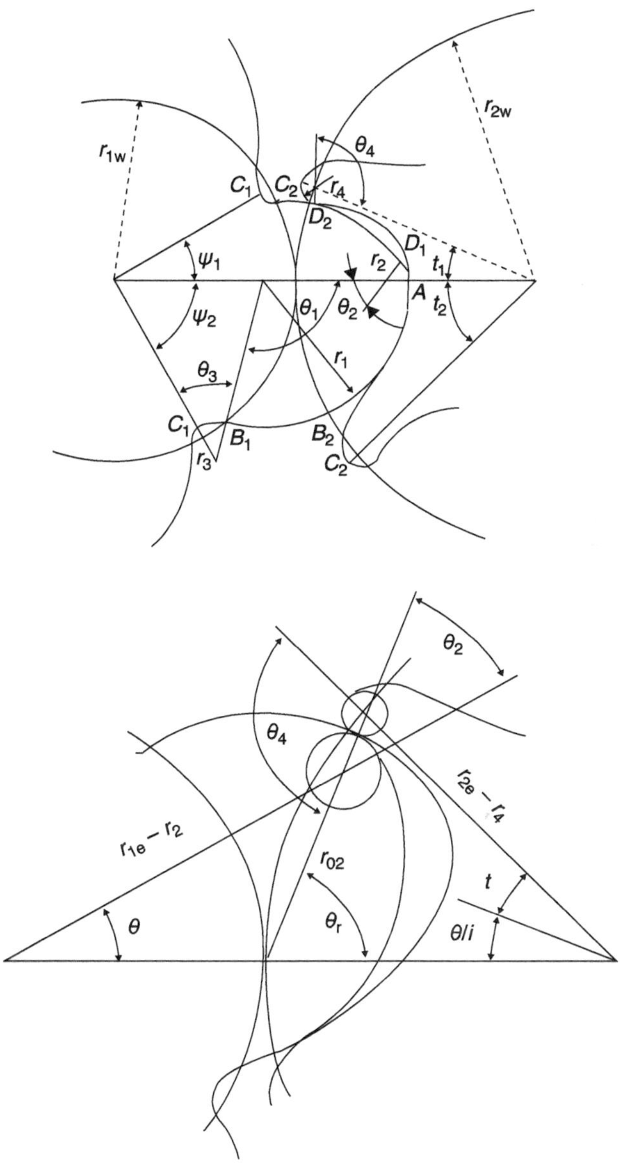

Figure 3.6 Demonstrator profile with its details

a variety of profiles can be generated, some of them with positive gate rotor torque, some of them suitable for low-pressure ratios, and some of them for high-pressure ratio expansion. The profile is fully computerized and can be used for demonstration, teaching and development purposes.

3.4.2 City 'N' rack generated rotor profile

The patented 'N' rotors are calculated by the rack generation procedure. This distinguishes 'N' rotors from almost any other type of screw expander rotor. The large blow-hole area, characteristic of the rack-generated rotors, is overcome in 'N' rotors by generating the high-pressure side of the rack by means of a rotor conjugate action which undercuts an appropriate curve on the rack. Such a rack is then used for profiling both the main and gate rotors. The method and its extensions were used to create a number of different rotor profiles.

The following is a brief description of a rack-generated 'N' rotor profile, typical for a family of rotor profiles designed for the efficient expansion of air, common refrigerants and a number of process gases. The rotors are generated by the combined rack–rotor generation procedure. Its features are such that it may be readily modified further to optimize performance for any specific application. The coordinates of all primary arcs on the rack are summarized here relative to the rack coordinate system. The lobe of the rack is divided into several arcs (Fig. 3.7).

All curves of the 'N' profile are given in a general arc form, from which circles, parabolae, ellipses and hyperbolae are easily generated. Segment D–E is a straight line on the rack, E–F is a circular arc of radius r_4, F–G is a straight line which generates the upper involute, while G–H on the rack is a meshing curve generated by the circular arc G_2–H_2 of radius

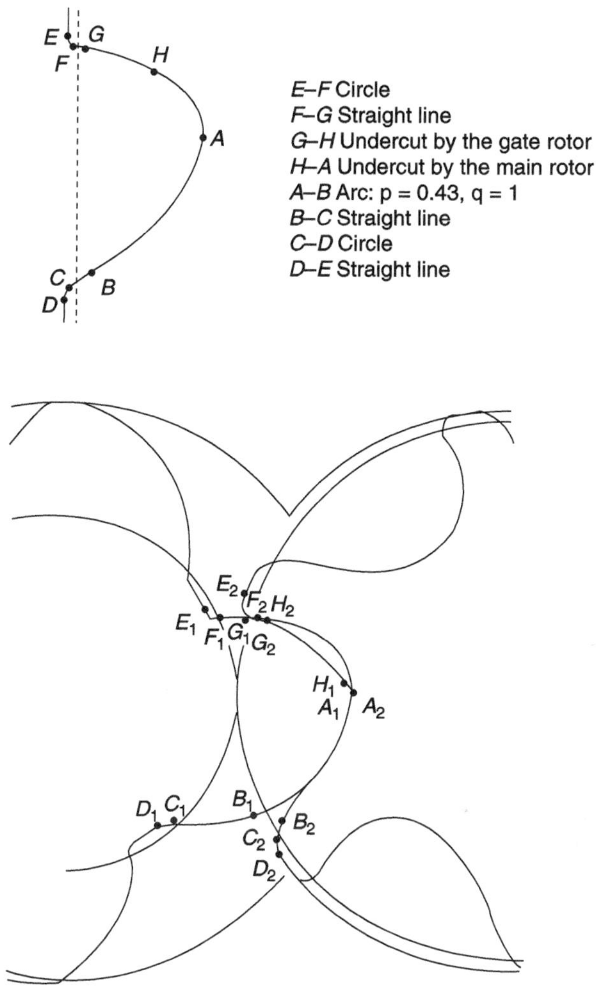

Figure 3.7 **City University 'N' profile details**

r_4 on the gate rotor. Segment *H–A* on the rack is a meshing curve generated by the circular arc H_1–A_1 of radius r_2 on the main rotor. Segment *A–B* is an arc that can be a circle, a parabola, a hyperbola or an ellipse. *B–C* is a straight line on the rack matching the involute on the rotor round lobe, and *C–D* is a circular arc on the rack of radius r_3.

Although the full evaluation of a rotor profile requires more than just a geometric assessment, some of the key features of the 'N' profile may be readily appreciated by comparing it with three popular screw rotor profiles, namely: 'Sigma', SRM 'D' and 'Cyclon', shown in Fig. 3.8. As can be seen, 'N' profiles have a greater throughput and a stiffer gate rotor for all cases, while other characteristics such as blow-hole area, confined volume and high-pressure sealing line lengths are identical in the designs compared. Also, the low-pressure sealing lines are shorter, but this is less important because the corresponding clearance can be kept small. The blow-hole area may be controlled by adjusting the tip radii on both the main and gate rotors and also by making the gate outer diameter equal to or less than the pitch diameter. Also, sealing lines can be kept very short by constructing

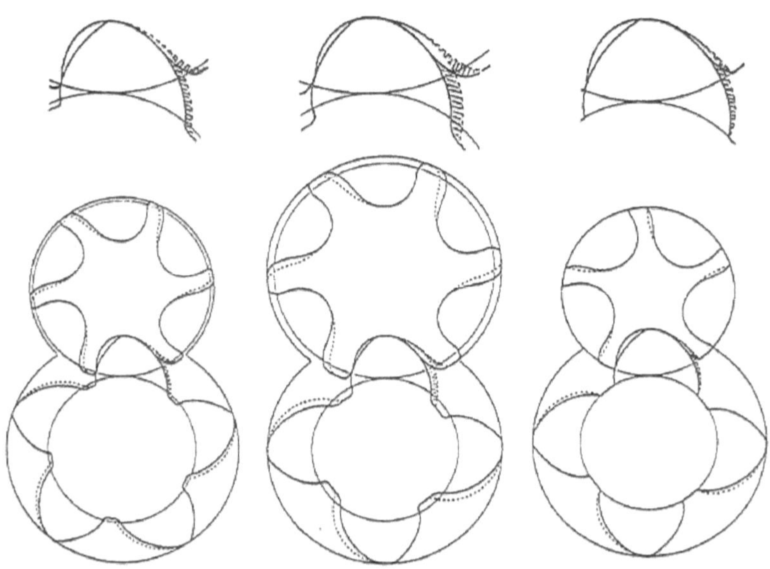

Figure 3.8 'N' rotors (solid lines) compared with Sigma (left), SRM 'D' (centre) and Cyclon (right) rotors, indicated by dotted lines

most of the rotor profile from circles whose centres are close to the pitch circle. However, any decrease in the blow-hole area will increase the length of the sealing line on the flat rotor side. A compromise between these trends is therefore required to obtain the best result.

3.4.3 Rotor position in expander bearings

A screw expander, especially the oil-flooded type, which operates with high-pressure differences, is heavily loaded by axial and radial forces that are transferred to the housing by the bearings. Rolling element bearings are normally chosen for small and medium screw expanders and these must be carefully selected to obtain a satisfactory design. Usually, two bearings are employed on the discharge end of each of the rotor shafts to absorb the radial and axial loads separately. Also, the distance between the rotor centre lines is in part determined by the bearing size and internal clearance. Any production imperfection in the bearing housing, such as displacement or eccentricity, will change the rotor position and thereby influence the expander behaviour.

The bearings, labelled 1–4, are presented in Fig. 3.9, together with their clearances and manufacturing tolerances of the bearing bores, δ_x and δ_y in the x and y direction respectively. The rotor centre distance is C, and the axial span between the bearings is a.

All imperfections in the manufacture of screw expander rotors should fall within and be accounted for by production tolerances. These are the wrong positions of the bearing bores, eccentricity of the rotor shafts, bearing clearances and imperfections, and rotor misalignment. Together, they account for the rotor shafts not being parallel. Let rotor movement δ_y in the y direction contain all displacements,

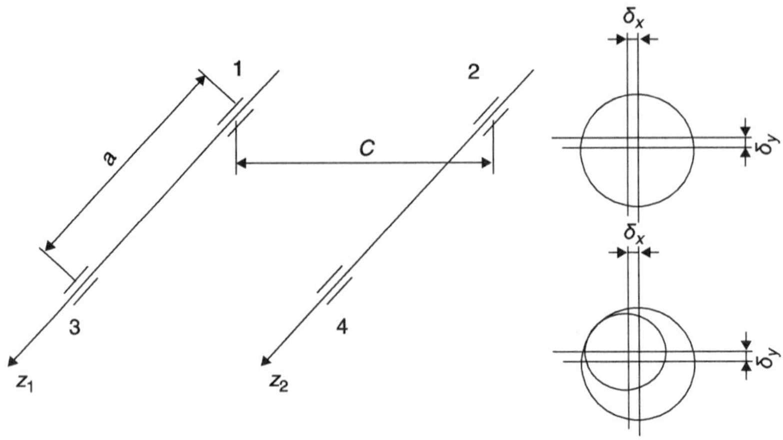

Figure 3.9 Rotor shafts in the expander housing and displacement in bearings (top), and coordinate systems of rotors with intersecting shafts (bottom)

and cause virtual rotation of the rotors around the x_1 and x_2 axes (Fig. 3.9). The movement δ_x can cause the rotor shafts to intersect. However, the movement δ_y causes the shafts to become non-parallel and non-intersecting. These both change the nature of the rotor position so that the shafts can no longer be regarded as parallel. The following analytical approach enables the rotor movement to be calculated and accounts for these changes.

Vectors $r_1 = [x_1, y_1, z_1]$ and r_2 now represent the helicoid surfaces of the main and gate rotors on intersecting shafts:

$$r_2 = [x_2, y_2, z_2] = [x_1 \cos\varsigma - z_1 \sin\varsigma - C, y_1, x_1 \sin\varsigma + z_1 \cos\varsigma] \qquad [3.14]$$

$$\tan\varsigma = \delta_x / a \qquad [3.15]$$

where ς is the rotation about the y axis. Since this rotation angle is usually very small, Eq. 3.14 can be rewritten in a simplified form as $r_2 = [x_2, y_2, z_2] = [x_1 - z_1\varsigma - C, y_1, x_1\varsigma + z_1]$.

The rotation ς will result in a displacement of $-z_1\varsigma$ in the x direction and a displacement of $x_1\varsigma$ in the z direction, while there is no displacement in the y direction. The displacement vector becomes $\Delta r_2 = [-z_1\varsigma, 0, x_1\varsigma]$.

In most practical cases, x_1 is small compared with z_1, and only displacement in the x direction needs to be considered. This means that rotation around the y axis will, in practice, only change the rotor centre distance. Displacement in the z direction may be significant for the dynamic behaviour of the rotors. Displacement in the z direction will be adjusted by the rotor relative rotation around the z axis, which can be accompanied by a significant angular acceleration. This may cause the rotors to lose contact at certain stages of the expander cycle and thus create rattling, which may increase the expander noise.

Since the rotation angle ς caused by displacement within the tolerance limits is very small, a two-dimensional presentation in the rotor end plane can be applied, as is done in the next section.

As shown in Fig. 3.5, in which the rotors on the non-parallel and non-intersecting axes are presented, vectors $r_1 = [x_1, y_1, z_1]$ and r_2, given by Eqs. 3.1 and 3.4, now represent the helicoid surfaces of the main and gate rotors on the intersecting shafts:

$$\begin{aligned} r_2 &= [x_2, y_2, z_2] \\ &= [x_1 - C, y_1 \cos\Sigma - z_1 \sin\Sigma, y_1 \sin\Sigma + z_1 \cos\Sigma] \end{aligned} \qquad [3.16]$$

$$\tan \Sigma = \delta_y / a \qquad [3.17]$$

Since angle Σ in Eq. 3.17 is very small, Eq. 3.16 can be rewritten in a simplified form as $r_2 = [x_2, y_2, z_2] = [x_1 - C, y_1 - z_1\Sigma, y_1\Sigma + z_1]$.

The rotation Σ will result in a displacement of $-z_1\Sigma$ in the y direction and a displacement of $y_1\Sigma$ in the z direction, while there is no displacement in the x direction. The displacement vector can be written as $\Delta r_2 = [0, -z_1\Sigma, y_1\Sigma]$.

Rotation about the x axis is effectively the same as if the main or gate rotor rotated relatively through angle $\theta = -z_1\Sigma/r_{1w}$ or $\tau = z_1\Sigma/r_{2w}$ respectively, and the rotor backlash will be reduced by $z_1\Sigma$. Such an approach substantially simplifies the analysis and allows the problem to be presented in two dimensions in the rotor end plane.

Although the rotor movements described here are entirely three-dimensional, their two-dimensional presentation in the rotor end plane section can be used for analysis. The end plane clearance gap can then be obtained from the normal clearance by appropriate transformation.

If δ is the normal clearance between the rotor helicoid surfaces, the cross product of the r derivatives, $\left(\frac{\partial r_1}{\partial t} \times \frac{\partial r_1}{\partial \theta}\right) \cdot \frac{\partial r_1}{\partial \tau}$, which defines the direction normal to the helicoids, can be used to calculate the coordinates of the rotor helicoids x_n and y_n from x and y to which the clearance was added as:

$$\begin{cases} x_n = x + p\dfrac{\delta}{D}\dfrac{\partial y}{\partial t}, y_n = y - p\dfrac{\delta}{D}\dfrac{\partial x}{\partial t} \\ z_n = \dfrac{\delta}{D}\left(x\dfrac{\partial x}{\partial t} + y\dfrac{\partial y}{\partial t}\right) \end{cases} \qquad [3.18]$$

where the denominator D is given as

$$D = \sqrt{p^2\left(\frac{\partial x}{\partial t}\right)^2 + p^2\left(\frac{\partial y}{\partial t}\right) + \left(x\frac{\partial x}{\partial t} + y\frac{\partial y}{\partial t}\right)^2} \qquad [3.19]$$

x_n and y_n serve to calculate new rotor end plane coordinates, x_{0n} and y_{0n}, with clearances from Eq. 3.18 for angles $\theta = z_n/p$ and τ respectively. x_{0n} and y_{0n} now serve to calculate the transverse clearance δ_0 as the difference between them, as well as the original rotor coordinates x_0 and y_0.

If, by any means, the rotors change their relative position, the clearance distribution at one end of the rotors may be reduced to zero on the flat side of the rotor lobes. In such a case, rotor contact will be prohibitively long on the flat side of the profile, where the dominant relative rotor motion is sliding. This indicates that the rotors will almost certainly seize in that region if they come into direct contact with one another. This situation indicates that a non-uniform clearance distribution should be applied to avoid hard rotor contact in rotor areas where sliding motion between the rotors is dominant.

3.4.4 Optimisation of rotor profiles

The full rotor and expander geometry, including the rotor throughput cross-section, rotor displacement, sealing lines and leakage flow cross-section, as well as the suction and discharge port coordinates, can be calculated from the rotor transverse plane coordinates and the rotor length and lead. The expander built-in volume ratio, which is determined by the inlet and discharge port sizes and shapes, is also an optimisation variable. As will be shown in the next chapter, these values are used as input parameters to calculate the expansion process with the aid of a thermodynamic model.

Once it is properly modelled, the expander geometry can easily be recalculated for any variation of the input parameters. Computation of the instantaneous cross-sectional area and working volume can thereby be calculated repetitively in terms of the rotation angle.

3.5 Geometry of rotor manufacturing tools

This section describes the generation of formed tools for screw expander hobbing, milling and grinding based on the envelope gearing procedure. The hobbing tool and milling/grinding tool are presented in Fig. 3.10.

3.5.1 Hobbing tools

Together, a screw expander rotor and its formed hobbing tool form a pair of meshing crossed helical gears with non-parallel and non-intersecting axes. The hob is simply a helical gear in which each tooth is referred to as a thread. Owing to their axes not being parallel, there is only point contact between the tool and the rotor whereas there is line contact between the two expander rotors. The need to satisfy the meshing equation (Eq. 3.11), leads to the rotor–hob meshing requirement for the given rotor transverse coordinate points x_{01} and y_{01} with their first derivative $\partial y_{01}/\partial x_{01}$. The hob

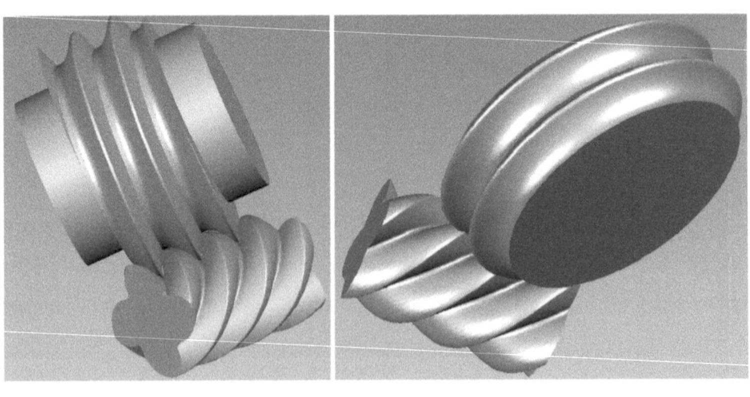

Figure 3.10 Rotor manufacturing tools: (left) hobbing tool, (right) milling/grinding tool

transverse coordinate points x_{02} and y_{02} can then be calculated. These are sufficient to obtain the coordinate $R_2 = \sqrt{x_{01}^2 + y_{01}^2}$. The axial coordinate z_2, calculated directly, and R_2 are hob axial plane coordinates, which define the hob geometry.

Reverse calculation of the hob–screw rotor transformation permits the transverse rotor profile coordinates to be determined, as a result of the manufacturing process. These may be compared with those originally specified to identify the effect of manufacturing errors, such as imperfect tool setting or tool and rotor deformation, upon the final rotor profile.

For the purpose of reverse transformation, the hob longitudinal plane coordinates R_2 and z_2, and $\partial R_2/\partial z_2$ should be given. The axial coordinate z_2 is used to calculate $\tau = z_2/p_2$, which is then used to calculate the hob transverse coordinates:

$$x_{02} = R_2 \cos\tau, \quad y_{02} = R_2 \sin\tau \qquad [3.20]$$

These are then used as given coordinates to produce the meshing criterion and the transverse plane coordinates of the manufactured rotors from Eq. 3.8.

3.5.2 Milling and grinding tools

Formed milling and grinding tools may also be generated by placing $p_2 = 0$ into the general meshing equation (Eq. 3.8) and then following the procedure of this section. The resulting meshing condition now is:

$$(C - x_1 + p_1 \cot\Sigma)\left(x_1 \frac{\partial x_1}{\partial t} + y_1 \frac{\partial y_1}{\partial t}\right) + p_1\left(p_1\theta \frac{\partial y_1}{\partial t} - C\cot\Sigma \frac{\partial x_1}{\partial t}\right) = 0 \qquad [3.21]$$

However, in this case, if one expects to obtain screw rotor coordinates from the tool coordinates, the singularity imposed does not permit the calculation of the tool transverse plane coordinates. The main meshing condition cannot therefore be applied. For this purpose, another condition is derived for the reverse milling tool to rotor transformation, from which the meshing angle τ is calculated:

$$\left(R_2 + z_2 \frac{dz_2}{dR_2}\right)\cos\tau + (p_1 + C\cot\Sigma)\frac{dz_2}{dR_2}\sin\tau + p_1\cot\Sigma - C = 0 \qquad [3.22]$$

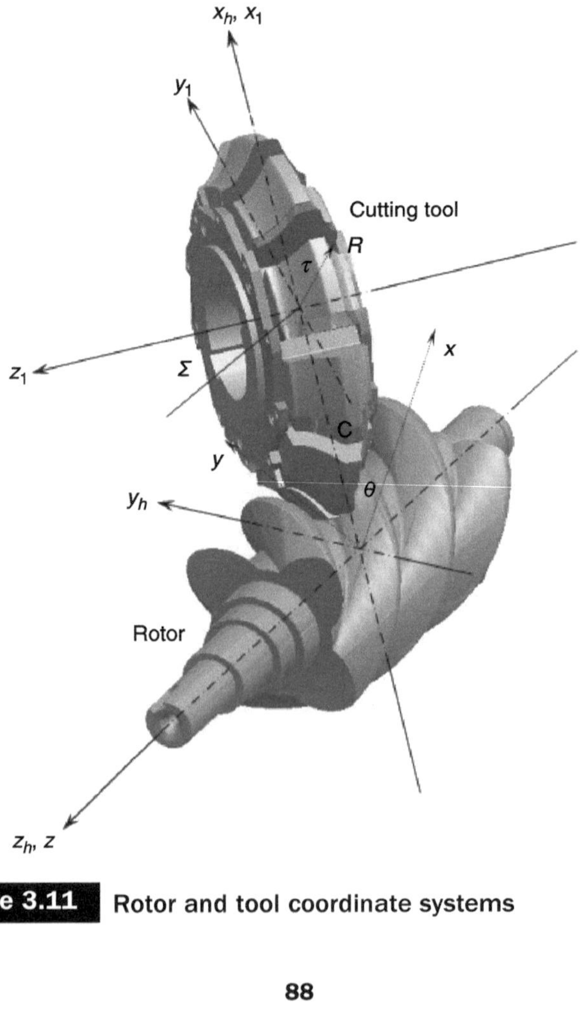

Figure 3.11 Rotor and tool coordinate systems

Once obtained, τ will serve to calculate the rotor coordinates after the manufacturing process. The rotor coordinates, thus obtained, will contain all the manufacturing imperfections, such as mismatch of the rotor–tool centre distance, error in the rotor–tool shaft angle, axial shift of the tool or tool deformation during the process as they are imposed on the calculation process.

The theory of profiling rotor manufacturing tools presented in this study is equally valid for both the main and gate rotor. A typical tool–rotor engagement is presented in Fig. 3.11 for the main rotor.

3.6 Design of screw expander housings and choice of bearings

Although advanced rotor profiles are a necessary condition for a screw expander to be efficient, all other components must be designed to take advantage of their potential if the full performance gains are to be achieved. Thus rotor to housing clearances, especially at the high-pressure end, must be properly selected.

The contact force between the rotors, which is determined by the torque transferred between them, plays a key role in expanders with direct rotor contact. The contact force is relatively small in the case of a main rotor-driven expander. In the case of a gate rotor drive, the contact force is substantially larger, and this case should be excluded from any serious consideration.

The same oil is used for oil flooding and for bearing lubrication, but the supply to and evacuation from the bearings are separate to minimize the friction losses. Oil is injected into the expander chamber at the place where thermodynamic calculations show the gas and oil inlet temperature to coincide. The position is defined on the rotor helix with the injection hole located so that the oil enters tangentially in line with the gate rotor tip in order to recover as much oil kinetic energy as possible.

To minimize the flow losses in the suction and discharge ports, the following features should be included. The suction port is positioned in the housing to let the gas enter with the fewest possible bends, and the gas approach velocity is kept low by making the flow area as large as possible. The discharge port size is first determined by estimating the built-in-volume ratio required for optimum thermodynamic performance. It is then increased in order to reduce the exit gas velocity and hence obtain the minimum combination of internal and discharge flow losses. Therefore, since the housing should be carefully dimensioned to minimise its weight, reinforcing bars across the discharge port are placed to improve its rigidity at higher pressures. The screw expander is a very simple machine with only two rotors and usually four to six bearings. It must be added that recent advances in the development of advanced low-friction rolling element bearings has contributed greatly to performance improvements.

Any production imperfection in the bearing housing, such as displacement or eccentricity, will change the rotor position and thereby influence the expander behaviour.

The trend is to obtain as small an expander as possible to produce and perform satisfactorily. This means that the rotor tip speed is kept as high as possible, but within some limits, determined by the conflicting trends of friction and leakage losses, resulting in acceptable expander efficiency. Rolling bearings should be used wherever possible since they permit smaller clearances than is possible with dynamic journal bearings, while the ports are made as wide as possible to

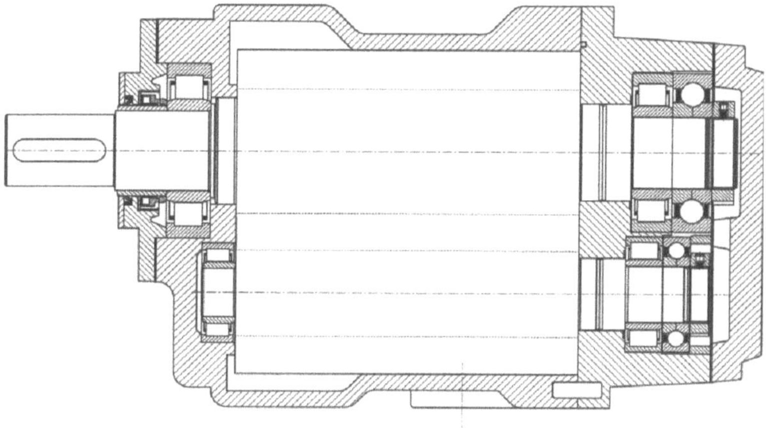

Figure 3.12 Drawing of typical screw rotors and housing assembled in a screw expander with low-pressure side bearings on the left and high-pressure side bearings on the right

minimise suction and discharge gas speeds. All these give very similar screw expander designs, and departures from them are almost negligible. An example of a screw expander layout is given in Fig. 3.12, where the rotors are presented inside of the housing, which is hatched.

4

Modelling and performance calculation of screw expanders

DOI: 10.1533/9781782421900.93

Abstract: This chapter presents a numerical model of the thermodynamic and fluid flow processes within screw machines which is valid for expander modes of operation. It includes the use of the equations of conservation of mass and energy applied to an instantaneous control volume of trapped fluid within the machine with allowance for fluid leakage, oil or other fluid injection, heat transfer and the assumption of real fluid properties. By simultaneous solution of these equations, pressure–volume diagrams may be derived of the entire expansion process within the machine. It is shown that performance predictions, based on this method of analysis, agree well with experimentally derived data. In view of the conflicting requirements of the large number of variables that affect expander performance, multi-variable optimization methods are required to obtain the best machine design.

Key words: screw expanders, mathematical modelling, thermodynamics, performance calculation, optimization.

Published by Woodhead Publishing Limited, 2014

4.1 The screw expander process and mathematical modelling

Screw machine analysis requires a combination of thermodynamic and fluid flow processes, both of which are dependent on the screw machine geometry. It is therefore rather complex and may be solved by the use of either one or multidimensional mathematical models. Both adiabatic and volumetric efficiencies are highly dependent on the rotor profile and the number of the lobes in each rotor.

The change of volume between each meshing pair of rotors, as a function of the rotation angle, is shown in Fig. 4.1, where the abscissa datum is set at zero at the inlet port cut-off point. As shown, the discharge port becomes

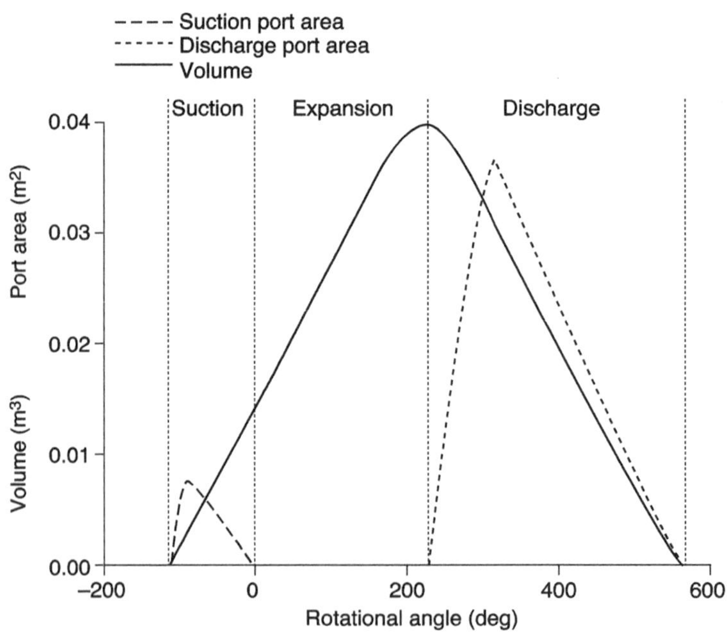

Figure 4.1 An example of volumetric change with rotation in a screw expander

exposed at the point of maximum volume. In practice, this point may be varied slightly with a negligible effect on performance.

Advances in mathematical modelling and computer simulation can be used to form a powerful tool for analysing and optimising the expander process. Such models have evolved greatly during the past ten years and, as they are better validated, their value as design tools has increased. Their use has led to a steady evolution in screw rotor profiles and compressor or expander shapes. The authors have developed a suite of subroutines for this purpose and their continued use should lead to further improvements in machine performance. The following sections describe the algorithms of the thermodynamics and flow processes used, which together comprise a set of equations that describe the physics of the complete process.

The mathematical model employs a differential kinematic relationship that describes the instantaneous operating volume and its change with rotation angle or time, together with the equations of conservation of mass and energy for that control volume. It also includes a number of algebraic equations, which define various phenomena specific to the suction, compression and discharge processes, as applicable. With all the described effects being accounted for in differential form, the model cannot be solved analytically. By this means, various simplifications of the equations, which have often been applied in the past to solve the equation set more efficiently, are easier to implement in this case and therefore unnecessary. It is therefore possible to use the model to observe the effects of neglecting some of the terms in the equations and to determine whether or not such simplifications are justified.

Solution of the equation set is performed numerically by employing the Runge-Kutta fourth-order method, with appropriate initial and boundary conditions.

The model accounts for a number of 'real-life' effects that may influence the final performance of a real machine. These provisions make the model suitable for a wide range of applications and include the following:

- The working fluid in the machine can be any gas or liquid–gas mixture with a known equation of state and derived equations for internal thermal energy and enthalpy, i.e. any ideal or real gas or liquid–gas mixture of known properties.
- The model has a provision to account for heat transfer between the gas and the machine screw or its casing; the model of heat exchange is specified in an approximate form, but still capable of reproducing reasonably well the real heat transfer effects.
- The model accounts for leakage of the working medium, which can occur at any stage of the process, through the clearances between the two rotors or between the rotors and stationary parts.
- The model works independently of the specification of machine geometry. Hence it will function for any input geometry.
- Any liquid, oil, water or refrigerant can be injected during the suction, expansion or discharge processes, for lubrication, sealing or cooling, and account taken of its effects on all the processes in the machine.
- The model also accounts for gas solubility in the injected fluid.
- The model includes thermodynamic equations of state and change of state of the fluid and their derivatives, which complete and close the equation set.

Certain assumptions had to be introduced in order to ensure efficient computation. These neither impose any specific limitations on the model, nor cause a significant departure from reality:

- The fluid flow in the model is assumed to be quasi-one-dimensional.
- Gas or gas–liquid mixture inflow or outflow through the machine suction or discharge ports is assumed to be isentropic.
- Leakage of the fluid through the clearances is assumed to be adiabatic.

4.2 Equations governing the screw machine processes

The working space (chamber) of a screw machine is an open thermodynamic system through which the mass flow varies with time. The control volume is the working chamber together with the suction and discharge plenums, for which the differential equations of conservation of energy and mass are written. These are derived using Reynolds' Transport Theorem.

A feature of the model is the use of the unsteady flow energy equation to compute the effect of volume change in a screw machine, within one cycle, on the thermodynamic properties of the working fluid. Internal energy rather than enthalpy is then the derived variable. This is advantageous when evaluating the properties of real fluids, because internal energy is independent of pressure, which is a derived property. Hence, by this means, the remaining thermodynamic and fluid properties at any point within the machine cycle are obtained without the need for iteration. However, computation still has to be carried out through several cycles until the solution converges.

The following forms of the conservation equations have been employed in the model.

The conservation of internal energy:

$$\omega\left(\frac{dU}{d\theta}\right)=\dot{m}_{in}h_{in}-\dot{m}_{out}h_{out}+Q-\omega p\frac{dV}{d\theta}$$

where θ is the angle of rotation of the main rotor, $h=h(\theta)$ is specific enthalpy, $\dot{m}=\dot{m}(\theta)$ is mass flow rate, $p=p(\theta)$ is fluid pressure in the working chamber control volume, $\dot{Q}=\dot{Q}(\theta)$ is heat transfer between the fluid and the machine surrounding, and $\dot{V}=\dot{V}(\theta)$ is the local volume of the machine working chamber. In the above equation, the subscript *in* denotes inflow and the subscript *out* the fluid outflow.

The fluid total inflow enthalpy consists of the following components:

$$\dot{m}_{in}h_{in}=\dot{m}_{suc}h_{suc}+\dot{m}_{lg}h_{lg}+\dot{m}_{oil}h_{oil}$$

where subscript *lg* denotes leakage gain, *suc* denotes suction conditions, and *oil* denotes oil.

The fluid total outflow enthalpy consists of:

$$\dot{m}_{out}h_{out}=\dot{m}_{dis}h_{dis}+\dot{m}_{ll}h_{ll}$$

where subscript *ll* denotes leakage loss, *dis* denotes the discharge conditions and $\dot{m}_{dis}$ denotes the discharge mass flow rate of the fluid contaminated with oil or other injected liquid.

The right-hand side of the energy equation consists of the following terms, which are modelled:

- The heat exchange between the fluid and the machine screw rotors and casing and the surroundings is accounted for by a heat transfer coefficient evaluated from the expression $\mathrm{Nu}=0.023\ \mathrm{Re}^{0.8}$. The characteristic length in the Reynolds and Nusselt number equations was taken as the difference between the outer and inner diameters of the main rotor. This may not be the most appropriate dimension for this purpose, but the characteristic length in the expression for the heat transfer coefficient has an exponent of 0.2 and therefore has little influence as long as it is of the same order of magnitude as other characteristic dimensions of the machine and as long as it characterises the machine size. The characteristic velocity for the Reynolds number is computed from the local mass flow and the cross-sectional area. Here, the surface over which the heat is exchanged, as well as the wall temperature, depends on the rotation angle θ of the main rotor.

- The energy gain due to the fluid inflow into the working volume is represented by the product of the mass intake and its averaged enthalpy. Thus, the energy inflow varies with the rotation angle. During the suction period, fluid enters the working volume with an averaged fluid enthalpy that dominates in the suction chamber. However, during the period when the suction port is closed, a certain amount of the expanded fluid leaks into the machine working chamber through the clearances. The mass of this fluid, as well as its enthalpy, is determined on the basis of gas leakage equations. The working volume is filled with fluid due to leakage only when the pressure in the space around the working volume is higher, otherwise there is no leakage, or it is in the opposite direction, i.e. from the working chamber towards other plenums.
- The total enthalpy inflow is further corrected by the amount of enthalpy brought into the working chamber by any injected fluid.
- The energy loss due to fluid outflow from the working volume is defined by the product of the mass outflow and its averaged fluid enthalpy. During delivery, this is the

expanded fluid entering the discharge plenum. The thermodynamic work supplied to the gas during the expansion process is represented by the term $p(\mathrm{d}V/\mathrm{d}\theta)$. This term is evaluated from the local pressure and local volume change rate. The latter is obtained from the relationships defining the screw kinematics, which yield the instantaneous working volume and its change with rotation angle. In fact, the term $\mathrm{d}V/\mathrm{d}\theta$ can be identified with the instantaneous interlobe area, corrected for the captured and overlapping areas.

- If oil or another liquid is injected into the working chamber of the machine, its mass inflow and enthalpy should be included in the inflow terms. Despite the fact that the injected liquid mass fraction in the mixture may be significant, its effect upon the volume flow rate is only marginal because its volume fraction is usually very small. The total fluid mass outflow also includes the injected fluid, the greater part of which remains mixed with the working fluid. Heat transfer between the gas and oil droplets is described by the first-order differential equation.

The mass continuity equation consists of:

$$\omega \frac{d\dot{m}}{d\theta} = \dot{m}_{in} h_{in} - \dot{m}_{out} h_{out}$$

The mass inflow rate consists of:

$$\dot{m}_{in} = \dot{m}_{suc} + \dot{m}_{lg} + \dot{m}_{oil}$$

And the mass outflow rate consists of:

$$\dot{m}_{out} = \dot{m}_{dis} + \dot{m}_{ll}$$

Each of the mass flow rates satisfies the continuity equation $\dot{m} = \rho w A$, where w (m/s) denotes fluid velocity, ρ is fluid density and A the flow cross-sectional area.

The instantaneous density $\rho = \rho(\theta)$ is obtained from the instantaneous mass m trapped in the control volume and the size of the corresponding instantaneous volume V as $\rho = m/V$.

4.3 Flow through the admission and discharge ports

The cross-sectional area A is obtained from the machine geometry and has been considered as a periodic function of the angle of rotation θ. The suction port area is defined by:

$$A_{suc} = A_{suc,0} \sin\left(\pi \frac{\theta}{\theta_{suc}}\right)$$

where subscript *suc* denotes the starting value of θ at the moment the suction port opens, and $A_{suc,0}$ denotes the maximum value of the suction port cross-sectional area. The reference value of the rotation angle θ is assumed to be at the point the suction port closes so that suction ends at $\theta = 0$, unless otherwise specified.

The discharge port area is similarly defined as:

$$A_{dis} = A_{dis,0} \sin\left(\pi \frac{\theta - \theta_c}{\theta_e - \theta_c}\right)$$

where subscript *e* denotes the end of discharge, *c* denotes the end of expansion and $A_{dis,0}$ is the maximum value of the discharge port cross-sectional area.

Suction and discharge port fluid velocities are defined by:

$$w = \mu\sqrt{2\left(h_2 - h_1\right)}$$

where μ is the suction/discharge orifice flow coefficient, while subscripts 1 and 2 denote the conditions downstream and upstream of the considered port respectively. Provision has been made in the computer code to account for reverse flow if $h_2 < h_1$.

4.4 Flow through the leakage paths

The leakages in a screw machine amount to a substantial part of the total flow rate and therefore are important because they influence the process both by affecting the machine mass flow rate or machine delivery, i.e. volumetric efficiency, and the thermodynamics of the expansion work efficiency. For practical computation of the leakage effects upon the machine process, it is convenient to distinguish two types of leakages, according to their direction. The gain leakages come from the suction plenum and from the neighbouring working chamber, which has a higher pressure. The loss leakages leave the chamber towards the discharge plenum and to the neighbouring chamber with a lower pressure.

Computation of the leakage velocity follows from approximate consideration of the fluid flow through the clearance. The process is essentially of the adiabatic Fanno-flow type. To simplify the computation, at some stages the conditions of constant temperature (T=const) or constant enthalpy (h=const) are applied. This departure from the prevailing adiabatic conditions has only a marginal influence if the analysis is carried out in differential form, i.e. for small changes in the rotational angle, as in the present model. The idealised clearance gap is assumed to have a rectangular shape and the mass flow of leaking fluid is expressed by the continuity equation:

$$\dot{m}_l = \mu_l \rho_l w_l A_g$$

where ρ and w are the density and velocity of the leaking gas, $A_g = l_g \delta_g$ is the clearance gap cross-sectional area, l_g is the leakage clearance length, δ_g is the leakage clearance width or gap, and $\mu = \mu(\mathrm{Re}, \mathrm{Ma})$ is the leakage flow discharge coefficient.

There are four different sealing lines in a screw machine, namely: the leading tip sealing line formed by the main and gate rotor forward tip and casing, the trailing tip sealing line formed by the main and gate reverse tip and casing, the front sealing line between the discharge rotor front and the housing, and the interlobe sealing line between the rotors. All sealing lines have clearance gaps that form leakage areas. Additionally, tip leakage areas must be added to the blow-hole areas.

Five different leakages can be identified according to the type and position of the leakage clearances. These are losses through the trailing tip sealing and front sealing and gains through the leading and front sealing. The fifth, 'through-leakage', does not directly affect the process in the working chamber, but passes through it from the suction port towards the discharge plenum.

Computation of the leakage fluid velocity follows from the momentum equation, which accounts for the fluid-wall friction:

$$w_l dw_l + \frac{dp}{\rho} + f\frac{w_l^2}{2}\frac{dx}{D_g} = 0$$

where $f(\mathrm{Re}, \mathrm{Ma})$ is the friction coefficient, which is dependent on the Reynolds and Mach numbers, D_g is the effective diameter of the clearance gap, $D_g \approx 2\delta_g$, and dx is the length increment. From the continuity equation, and assuming that $T = \mathrm{const}$, to eliminate gas density in terms of pressure, the equation can be integrated in terms of pressure from the high-pressure side at position 2 to the low-pressure side at position 1 of the gap to yield:

$$\dot{m}_l = \rho_l w_l A_g = \sqrt{\frac{p_2^2 - p_1^2}{a^2\left(\zeta + 2\ln\frac{p_2}{p_1}\right)}}$$

where $\zeta = f\ L_g/D_g + \Sigma\xi$ characterises the leakage flow resistance, with L_g the clearance length in the leakage flow direction and f the friction factor and local resistance coefficient. ζ can be evaluated for each clearance gap as a function of its dimensions and shape and flow characteristics. a is the speed of sound. The full procedure requires the model to include the friction and drag coefficients in terms of Reynolds and Mach numbers for each type of clearance.

Similarly, the working fluid friction losses can also be defined in terms of the local friction factor and fluid velocity related to the tip speed, density and elementary friction area. The model uses a value of ζ in terms of a simple function for each particular machine type and employs it as an input parameter.

These equations are incorporated into the model of the machine and employed to compute the leakage flow rate for each clearance gap at the local rotation angle θ.

4.5 Injection of oil and other liquids

Injection of oil or other liquids for lubrication, heating, cooling or sealing purposes modifies the thermodynamic process in a screw machine substantially. The following paragraph outlines the procedure for accounting for the effects of oil injection. The same procedure can be applied to treat the injection of any other liquid. Special effects, such as vapour or its condensate mixing and resolving in the injected fluid or vice versa, should be accounted for separately if they are expected to affect the process. A procedure for incorporating these phenomena into the model will be outlined later.

A convenient parameter to define the injected oil mass flow is the oil-to-gas mass ratio, m_{oil}/m_{gas}, from which the oil inflow through the open oil port, assumed to be uniformly distributed, can be evaluated as:

$$\dot{m}_{oil} = \frac{\dot{m}_{oil}}{\dot{m}_{gas}} \dot{m} \frac{z_1}{2\pi}$$

where the oil-to-gas mass ratio is specified in advance as an input parameter.

In addition to lubrication, a major purpose for injecting oil into a machine is to heat or cool the working fluid. To enhance the efficiency of these processes, the oil is atomised into a spray of fine droplets in order to increase the contact surface between the working fluid and the oil. The atomisation is performed by using specially designed nozzles or by simple high-pressure injection. The distribution of droplets sizes can be defined in terms of the oil-working fluid mass flow and velocity ratio for a given oil-injection system. The path of each distinct class of oil droplets can be followed, until they hit the rotor or casing wall, by solving the dynamic equation for each droplet size in a Lagrangian frame, accounting for inertial, gravity, drag and other forces. The solution of the droplet energy equation, in parallel with the momentum equation, should yield the amount of heat exchange with the surrounding working fluid.

In the present model, a simpler procedure for the heat exchange with the working fluid was determined from the differential equation for instantaneous heat transfer between the surrounding gas and an oil droplet. Assuming that the droplets retain a spherical form, with a prescribed Sauter mean droplet diameter d_S, the heat transfer between the droplet and the working fluid can be expressed in terms of the simple cooling law $Q_o = h_o A_o\ (T_{gas} - T_{oil})$, where A_o is the droplet surface, $A_o = d_S^2 \pi$, d_S is the mean Sauter diameter of the droplet and h_o is the heat transfer coefficient at the droplet surface, determined from the empirical expression $\mathrm{Nu} = 2 + 0.6\mathrm{Re}^{0.6}\mathrm{Pr}^{0.33}$. The exchanged heat must balance the rate of change of heat taken or given away by the droplet in unit time, given by $Q_o = m_o c_{oil} dT_o/dt = m_o c_{oil} \omega dT_o/d\theta$, where c_{oil} is the oil specific heat and the subscript o denotes oil droplet. The rate of change of oil droplet temperature can now be expressed as:

$$\frac{dT_o}{d\theta} = \frac{h_o A_o \left(T_{gas} - T_o\right)}{\omega m_o c_{oil}}$$

The heat transfer coefficient h_o is obtained from:

$$\mathrm{Nu} = 2 + 0.6\mathrm{Re}^{0.6}\mathrm{Pr}^{0.33}$$

Integration of the equation in two time/angle steps yields the new oil droplet temperature at each new time/angle step:

$$T_o = \frac{T_{gas} - kT_{o,p}}{1 + k}$$

where $T_{o,p}$ is the oil droplet temperature at the previous time step and k is the non-dimensional time constant of the droplet, $k = \tau/\Delta t = \omega\tau/\Delta\theta$ with $\tau = m_o c_{oil}/h_o A_o$ being the real time constant of the droplet. For the given mean Sauter diameter d_S, the non-dimensional time constant takes the form:

$$k = \frac{\omega m_o c_{oil}}{h_o A_o \Delta\theta} = \frac{\omega d_S c_{oil} \rho}{6 h_o \Delta\theta}$$

The droplet temperature, thus obtained, is further assumed to represent the average temperature of the oil (i.e. $T_{oil} \approx T_o$), which is further used to compute the enthalpy of the gas–oil mixture.

This approach is based on the assumption that the oil-droplet time constant τ is smaller than the droplet travelling time through the gas before it hits the rotor or casing wall, or reaches the machine discharge port. This means that the heat exchange is completed within the droplet travelling time through the working fluid during expansion. This prerequisite is fulfilled by atomisation of the injected oil to produce sufficiently small droplet sizes to give a small droplet time constant, as well as by choosing an adequate nozzle angle and, to some extent, the initial oil spray velocity. Separate computation of the droplet trajectory, based on the solution of the droplet momentum equation for different droplet mean diameters and initial velocities, indicates that for most screw machines currently in use, except, perhaps, for the smallest ones, with typical tip speed velocities between 20 and 50 m/s, this condition is well satisfied for oil droplets with diameters of less than 50 μm.

Since the inclusion of a complete model of droplet dynamics would complicate the computer code unnecessarily and the outcome will always be dependent on the design and angle of the oil injection nozzle, the present computation code uses the above-described simplified approach, which was found to be fully satisfactory for a range of different machines. The input parameter is only the Sauter mean diameter of the oil droplets, d_S, and the oil density, viscosity and specific heat.

4.6 Solution procedure for the screw machine thermodynamics

To summarise, the description of the thermodynamic processes in a screw machine is completed by differential equations for the lobe volumes, which define $V(\theta)$ and $dV/d\theta$, by the differential equation of internal thermal energy, and by the differential equations describing the working chamber mass balance. Boundary pressures and temperatures in the suction and discharge chambers are known. In addition, the algebraic equations of state and specific internal energy and specific enthalpy are sufficient to obtain the mass flows through the suction and discharge ports and through the clearances, the mass in the working chamber, the pressure and temperature of the fluid in the working chamber, and the mass and the temperature of the injected oil.

If the fluid states, described by the pressure and temperature in the pressure and suction plenums, are considered to vary with the rotation angle, it is necessary to couple the differential equations for energy and mass flow rates. The total number of differential equations is then increased by another two for each plenum.

All the differential equations are solved by means of the Runge-Kutta fourth-order procedure. As the initial conditions were arbitrarily selected, the convergence of the solution is achieved by applying cyclic boundary conditions when the difference between the two consecutive machine cycles reaches a sufficiently small monitoring value prescribed in advance.

The instantaneous bulk density ρ is obtained from the instantaneous mass trapped in the control volume and the size of the corresponding instantaneous volume V as $\rho = m/V$.

The equations of energy and continuity are solved to obtain $U(\theta)$ and $m(\theta)$. With $V(\theta)$, the specific internal energy and the specific volume $u = U/m$ and $v = V/m$ known, T and p or x can then be calculated.

For an ideal gas, the internal thermal energy of the gas–oil mixture is given by:

$$U = (mu)_{gas} + (mu)_{oil} = \frac{mRT_{gas}}{\gamma - 1} + (mcT)_{oil} = \frac{pV}{\gamma - 1} + (mcT)_{oil}$$

where R is the gas constant and γ the adiabatic exponent. Hence, the pressure or temperature of the fluid in the machine working chamber can be explicitly calculated with the aid of the equation for the oil temperature T_{oil}:

$$T = (\gamma - 1)\frac{(1+k)U - (mcT)_{oil}}{(1+k)mR + (mc)_{oil}}$$

If k tends 0 (i.e. for high heat transfer coefficients or small oil droplet size), the oil temperature approaches the gas temperature rapidly. In that case, T and p are calculated explicitly.

For the case of a real gas, the situation is more complex, because the temperature and pressure cannot be calculated explicitly. However, since the internal energy can be expressed as a function of the temperature and specific volume only, the calculation procedure can be simplified by employing the internal energy as a dependent variable instead of enthalpy, as often is the practice. The equation of state $p = f_1(T, V)$ and the equation for specific internal energy $u = f_2(T, V)$ are usually decoupled. Hence, the temperature can be calculated from the known specific internal energy and the specific volume obtained from the solution of differential equations, while the pressure can be calculated explicitly from the temperature and the specific volume by means of the equation of state.

In the case of the phase change of a wet vapour during the expansion process, the specific internal energy and volume of the liquid–gas mixture are:

$$u = (1 - x)u_f + xu_g \qquad v = (1 - x)v_f + xv_g$$

where u_f, u_g, v_f and v_g are the specific internal energy and volume of liquid and gas and are functions of the saturation temperature only. The equations require an implicit numerical procedure that is usually incorporated in property packages. As a result, temperature T and dryness fraction x are obtained. These equations are in the same form for any kind of fluid, and they are essentially simpler than any others in their derived form. In addition, the inclusion of any additional phenomena into the differential equations of internal energy and continuity is straightforward.

4.7 Calculation of thermodynamic properties of working fluids

A description of the thermodynamic properties of pure fluids and their mixtures is provided in Chapter 1, with further details given in Appendix 1 and Appendix 2.

4.8 Calculation of machine performance parameters

Numerical solution of the mathematical model of the physical process in the machine provides a basis for a more exact computation of all desired integral (bulk) characteristics with a satisfactory degree of accuracy, and in that respect is superior to the more empirical integral approach. The most important of these properties are the machine mass flow rate $\dot{m}$ (kg/s), the indicated power P_{ind} (kW), the volumetric efficiency (η_v), the adiabatic efficiency (η_a), the isothermal efficiency (η_t) and the indicated efficiency (η_i).

The instant fluid mass trapped in the working volume is determined as the difference between the total fluid mass inflow and outflow:

$$m = m_{in} - m_{out}$$

where m_{in} and m_{out} are obtained from the integration of the corresponding differential equations over the cycle. During the rotation of the machine shafts, due to the difference between the main and gate rotor shaft speeds, only the number of working volumes of the main screw lobes z_1 contributes towards the process, so that the total mass participating is mz_1. Hence the actual fluid mass flow $\dot{m}$ (kg/s) is

$$\dot{m} = mz_1 n/60$$

where n is the number of revolutions per minute of the main rotor.

The volume delivery $\dot{V}$ is defined with reference to the suction conditions and is usually expressed in m^3/min:

$$\dot{V} = 60\,\mathrm{m}/\rho_0$$

From the known maximum volume of the working chamber, the theoretical mass flow is:

$$\dot{m}_t = \frac{\left(F_{1n} + F_{2n}\right) L n z_1 \rho}{60}$$

where $F_1 n$ and $F_2 n$ are the cross-sectional areas of the lobes in the front plane of the main screw and the gate screw, respectively, and L is the length of the screw. The volumetric efficiency is readily obtained as:

$$\eta_v = \frac{\dot{m}}{\dot{m}_t}$$

It is worth noting that the effects of leakage, working fluid heat exchange and the gas retention in the pockets on the pressure side of asymmetric lobe profiles, are all included in the volumetric coefficient through the differential treatment of the governing equations of the mathematical model.

The indicated work transferred to the screw rotors during the suction, expansion and discharge processes is represented by the area of the indicated p–V diagram:

$$W_{ind} = \int_{cycle} V\,dp$$

Within the indicated work, flow losses during suction, expansion and discharge, leakages and heat exchange, as well as the influence of injected oil are included into the differential equations of the model in the same way.

The indicated work in a single machine working chamber is further used for the computation of the machine indicated power:

$$P_{ind} = \frac{W_{ind} z_1 n}{60}$$

In addition to the indicated power, it is useful to know the specific indicated power:

$$W_{sind} = \int_{cycle} \frac{V}{m} dp$$

where m is the mass of the fluid contained in the working chamber V at the considered instant of time.

The indicated work can be compared with the theoretical adiabatic or isothermal work to yield the corresponding efficiency:

$$\eta_t = \frac{W_{ind}}{W_t} \qquad \eta_a = \frac{W_{ind}}{W_a}$$

Here the theoretical isothermal work and adiabatic work are determined from the common theoretical expressions. For an ideal gas, the theoretical isothermal and adiabatic works for ideal gas are, respectively:

$$W_t = RT_1 \ln\frac{p_2}{p_1} \qquad W_a = \frac{\gamma}{\gamma - 1} R(T_1 - T_2)$$

where 1 denotes the beginning and 2 the end conditions of the expansion process.

4.9 Results of modelling and experimental investigations

4.9.1 The pressure–volume diagram

The most distinctive feature of the derived results is the shape of the pressure–volume diagram of which a typical result is shown in Fig. 4.2 for a two-phase expansion. As can be seen, the initial filling process, which is normally assumed to take place at approximately constant pressure, is in fact associated with a relatively large pressure drop and hence a significant expansion. It is radically different from normal compressor test results in this region. This pressure drop is due to the acceleration of the fluid as it enters the varying and restricted space between the rotor lobes. It is much larger than the pressure change associated with gas discharge from compressors through the same area in the reverse direction because the fluid entering has a liquid content and therefore a much greater density. For the same fluid velocity, the ρV^2 value is therefore much bigger. It is worth noting that a similar but less marked effect was recently noted when investigating oil injection in a dry gas screw expander. The gain in fluid kinetic energy is not dissipated in the filling process, as in reciprocating machines, because the entire trapped volume of fluid rotates. The filling process is therefore the non-steady flow equivalent of the steady flow through a turbine nozzle.

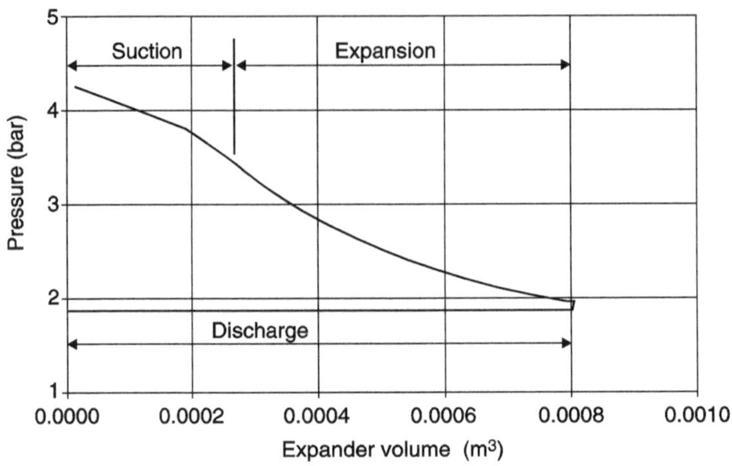

Figure 4.2 **A typical estimated *p–V* diagram of a two-phase expansion process. Note the large pressure drop associated with the filling process**

The reliability of our predictions is demonstrated in Figs. 4.3 and 4.4, where a representative set of the *p–V* measurements is compared with their corresponding computer simulations for the same inlet conditions, rotational speed and back pressure. These cover both low- and high-speed tests with over- and under-expansion in the rotors.

The implications of our findings are as follows:

1. Especially in two-phase expansion, the pressure ratio of expansion is much larger than that anticipated from the built-in volume ratio of the expander and increases with speed for the same inlet conditions.
2. Given the same inlet conditions, the greater the rotational speed, the more fluid flashes into vapour during the filling process. Hence the gain in mass flow rate decreases.

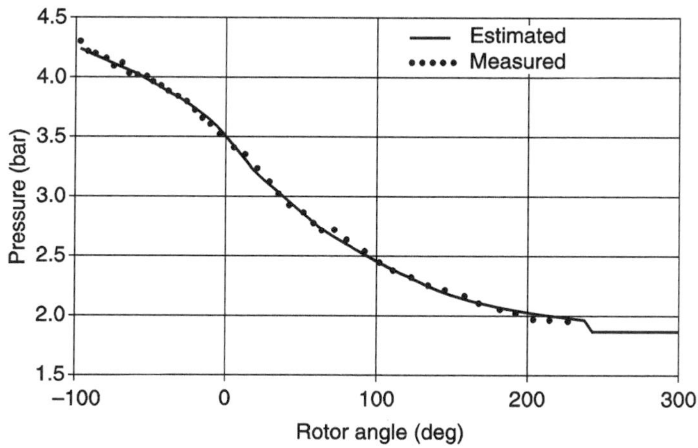

Figure 4.3 Predicted and measured pressure change with rotation in a 163 mm diameter rotor screw expander operating at 10 m/s tip speed with R113 as the working fluid

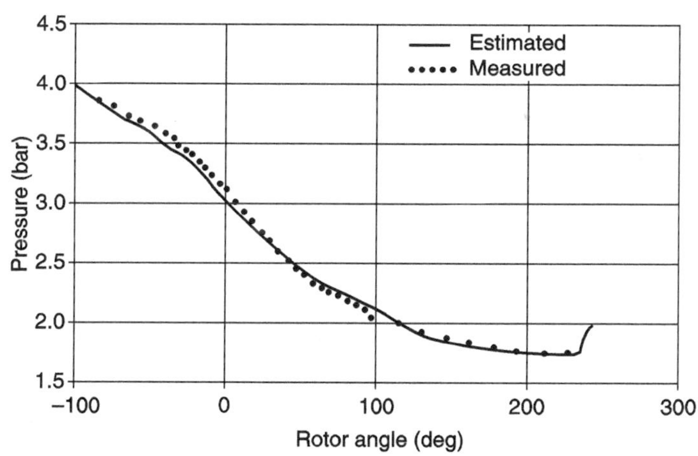

Figure 4.4 Predicted and measured pressure change with rotation in a 163 mm diameter rotor screw expander operating at 20 m/s tip speed with R113 as the working fluid. Speed = 3600 rev/min and inlet temperature = 100 °C

4.9.2 Accounting for mechanical friction losses

Numerical solution of the differential equations described in this chapter enables the mass flow to be estimated and both the indicated power output and adiabatic efficiency to be derived from the area under the p–V diagram. The measured fluid flow rate may be compared directly with its estimated value but the power output and adiabatic efficiency of the expander obtained from measurements is not directly comparable with their corresponding estimated values unless the friction losses associated with the timing gear, bearings and shaft seals are accounted for. These losses vary widely depending on the type of bearings and seals used in each design. Accordingly, the method used to estimate them in each expander was to compare the calculated indicated power with the measured shaft output at a test point where the estimated and measured mass flows agreed fairly well. The difference between these values was taken as the cumulative friction loss, which was assumed to be proportional to rotational speed squared and independent of load. The power loss function so derived was then used to predict the values of shaft output and adiabatic efficiency at all the other test points.

4.10 Calculation of pressure forces acting on screw machine rotors

Screw machine rotors are subjected to severe pressure loads. The rotors, as well as their bearings, must satisfy rigidity and elasticity requirements to ensure appropriate and reliable machine operation.

As shown in Fig. 4.5, in position (a) there is no contact between the rotors. Since A and B are on the circle, overall

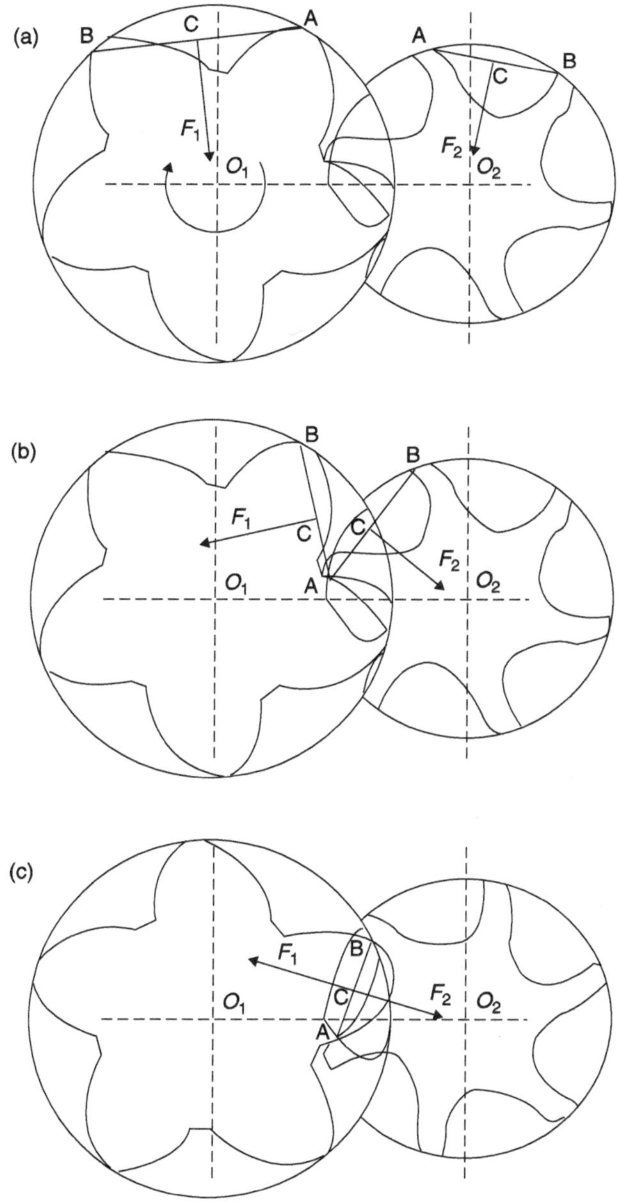

Figure 4.5 Pressure forces acting on screw machine rotors

forces F_1 and F_2 act towards rotor axes and there are radial forces only. There is no torque caused by pressure forces in this position.

In position (b), there is only one contact point between the rotors, at point A. Forces F_1 and F_2 are eccentric, and have both radial and circumferential components. The latter cause the torque. Due to the force position, the torque on the gate rotor is significantly smaller than that on the main rotor.

To appreciate the calculation of the pressure loads, several cases are considered. Let the pressure $p(\theta)$ be known for any instantaneous angle of rotation θ, with a reasonable angle increment of, say, 1°. Figure 4.5 presents the radial and torque forces at a rotor cross-section. The example is given for 5/6 'N' rotors. Pressure p acts in the corresponding interlobes normal to line AB. A and B are either on the sealing line between the rotors or on the rotor tips. Since they lie on the sealing line, they are fully defined by the rotor geometry.

In position (c), both contact points are on the rotors, the overall and radial forces are equal for both rotors. As in the previous case, they also cause the torque.

4.11 Radial, axial rotor loads, torque and bearing reactions

Let the x direction be parallel to the line between rotor axes O_1 and O_2. y is perpendicular to x. The radial force components are:

$$R_x = -p\int_A^B dy = -p\left(y_B - y_A\right)$$

$$R_y = -p\int_A^B dx = -p\left(x_B - x_A\right)$$

The torque is:

$$T = p\int_A^B xdx + p\int_A^B ydy = 0.5p\left(x_B^2 - x_A^2 + y_B^2 - y_A^2\right)$$

The above equations are integrated along the profile for all profile points. Then they are integrated for all angle steps to complete one revolution employing a given pressure history $p = p(\theta)$. Finally, the sum for all rotor interlobes is made taking into account the phase shift, as well as the axial shift between the interlobes.

Since the gate rotor has a lead angle proportional to the gearing ratio z_2/z_1 and is larger than that of the main one, where z are the rotor lobe numbers, appropriate summation usually leads to the larger radial forces on the gate rotor despite the fact that it may be smaller than the main rotor. The axial force is a product of the pressure and interlobe cross-section. In some regions, the interlobes overlap each other. The gate rotor covers a small part of the main rotor interlobe, while the main rotor covers the majority of the gate interlobe. This phenomenon causes the main axial force to be disproportionately larger than the gate axial force. A correction is allowed for axial forces, which takes into account the fact that the pressure in the rotor front gaps also acts in the axial direction, by using the average of pressures in two neighbouring interlobes to act on the lobe in question.

Rotor axial forces act to minimize the discharge bearing radial force increasing the suction bearing forces. This is generally convenient, because the suction bearing forces are usually smaller than the discharge forces. Bearing in mind that the main rotor axial force is larger than that of the gate rotor, this effect is more beneficial for the main rotor.

Since the radial force R and its axial position z_1 are calculated in advance for every rotation angle θ, δ calculated from the expression above is also a function of the rotation angle.

4.12 Rotor deflections

If the rotors are loaded by radial force R with bearing reactions R_D and R_S on the discharge and suction rotor sides over a span z_2, the rotor elastic line function is given by the differential equation:

$$\frac{d^2\delta}{dz^2} = \frac{M}{EI}$$

where $\delta = \delta(z)$ is the rotor bending deflection, M is the bending moment function, E is the modulus of elasticity and I is the rotor polar moment of inertia, calculated from the rotor geometry, by means of numerical integration.

Integration of the above equation over the rotor span between the two radial bearings gives the bending deflection as a function of the rotor axial coordinate z, which has its own maximum value. This is calculated for every increment of the rotation angle.

4.13 Recent advances in screw machine development

Since modern manufacturing methods allow for much tighter clearances, internal machine leakage rates have been reduced. Hence, further improvements in screw machine design are possible only by using more refined analytical principles that can quantify the effects on performance of each of the controlling design variables. Accordingly, improved mathematical modelling to account for all of them is now essential and further improvements in design are now dependent on them.

The main requirement is to improve the rotor profiles so that the internal flow area through the machine is maximised while the leakage path is minimised and internal friction due to relative motion between the contacting rotor surfaces is made as small as possible. This is achieved through several steps described below.

4.13.1 Proper choice of rotor configuration

Increasing the number of rotor lobes enables the same built-in volume ratio to be attained with larger suction ports. Larger suction ports decrease the inlet velocity and therefore reduce the suction pressure drop, thereby reducing the possibility of over-expansion and consequent reduction in the machine overall efficiency. The greater the number of lobes, the smaller the pressure difference between the two neighbouring working chambers. Thus, interlobe leakage losses are reduced. Furthermore, more lobes combined with a large wrap angle ensure multiple rotor contacts, which reduce vibrations and thus minimize noise. However, more lobes usually mean less rotor throughput, which implies that rotors with more lobes are somewhat larger than their counterparts with fewer lobes. Also, the leakage-to-delivery ratio is worse with more rotor lobes. Therefore, such machines are less efficient. Additionally, more lobes increase the manufacturing cost.

4.13.2 Considerations of rotor sealing line length and blow-hole area

Since screw machines tend to rotate relatively slowly, rotor profiles must have the smallest possible blow-hole area if leakages are to be minimised. However, reduction of the blow-hole area is associated with an increase in the sealing line length. It is therefore necessary to find the optimum profile shape that minimises the sum of both the blow-hole and sealing line leakage areas.

4.13.3 Rotor proportions

A general feature of screw machines is that the pressure difference through them causes high rotor loads, especially for low-temperature refrigeration machines, where these are large. Therefore, to maintain their rigidity and minimise deflection, rotor profiles usually have a relatively small male rotor addendum in order to increase the female root diameter. This sometimes leads to very shallow and clumsy rotors. An alternative possibility is to increase the female rotor lobe thickness. This greatly increases its moment of inertia and thereby reduces the rotor deflection more effectively.

4.13.4 Choice of rotor wrap angle

Increasing the rotor wrap angle is generally associated with reducing the interlobe sealing line and hence with reduced leakage between the rotors. Contemporary trends in screw machine design are therefore towards larger wrap angles. However, on occasion, this has led to exceeding the limiting values and thereby reducing the machine displacement.

4.13.5 Progress in machine bearings

In some machine designs, multiple cylinder roller bearings or multipoint ball bearings are located at the high-pressure end of the rotors to withstand the large radial forces reliably over a long operating life. Frequently, two bearings are also employed for axial loads. Since only one axial bearing actually takes the load, the role of the other is mainly to prevent rotor bounce in the axial direction.

4.13.6 Rotor clearance distribution and contact at the flat lobe side

Oil-flooded machines have direct contact between their rotors. In well-designed rotors, the clearance distribution will be set so that this is first made along their so-called contact bands, which are positioned close to the rotor pitch circles. Since the relative motion between the contacting lobes in this region is almost pure rolling, the danger of them seizing, as a result of sliding contact, is thereby minimised. The traditional approach is to maintain a high 'positive gate rotor torque', which ensures round flank contact. What is not widely appreciated is that there are significant advantages to be gained by maintaining a negative gate rotor torque to ensure that contact, when it occurs, will be on the flat lobe face. The reason for this can be understood by examination of the sealing line lengths. That on the flat flank is much longer than that of the round flank. Thus, minimising the clearance on the flat flank will reduce the interlobe leakage more than minimising the round flank clearance. Also, negative gate torque is achieved by making the gate rotor lobes thicker and the main rotor lobes correspondingly thinner. The displacement is thereby increased. Thus both these effects lead to higher machine flows and efficiencies.

4.13.7 Account of thermal expansion of the rotors and housing

Although the temperature range over which screw machines operate is not large, the effects of thermal expansion are highly significant if the small clearances required between the rotors and between the rotors and the housing are to be maintained under working conditions. Thus, the rotor clearances obtained under manufacturing conditions must be estimated while taking account of the thermal distortion that occurs when the machine reaches its operating temperature and pressure, and calculation must allow for unequal expansion of the rotors in different coordinate directions.

4.13.8 Introduction of bearing centre displacement

An additional design aspect, which though important is not widely appreciated, is that the pressure loads will tend to push the rotors apart from their design position in the casing, as a result of the clearances within the bearings. If these are not taken fully into account, the resulting displacement will cause contact between the rotor tips and the casing, when the rotor clearances are small and the pressure loads are high. To counter this, the bearing centre distance must be smaller than that of the rotor housing bores. To keep the rotor interlobe clearance as small as possible, the bearing centre distance must be even further reduced.

Also, if the bearing centres are set to be the same as those of the rotors, the clearance between the rotors and housing will be smaller at the low-pressure side of the rotors and larger at the high-pressure side. Since leakage is caused by the pressure difference, this displacement creates the least favourable rotor position for efficient machine operation. The bearing centre distances must therefore be arranged to maintain a uniform clearance between the rotors and the housing.

4.13.9 Optimisation of the machine process

Analysis of machine behaviour shows that there are conflicting requirements for desirable machine characteristics. This implies that only simultaneous optimisation of all the variables involved in the design process will lead to the best possible machine performance. A full multivariable optimisation of screw machine geometry and operating conditions should be performed to establish the most efficient machine design for a given duty. This can be achieved by using a computer software package, based on a Box constraint simplex method, which provides the general specification of the rotor and machine characteristics in terms of several key parameters and which can generate various rotor and machine shapes.

5

Applications for screw expanders

DOI: 10.1533/9781782421900.127

Abstract: This chapter begins with a more detailed description of problems associated with the lubrication of screw expanders and some solutions to them. It then differentiates between applications of screw expanders, in systems in which the fluid, constituting the source of energy, is expanded to produce power within a larger system, which may have other functions and cycles, where the sole aim of the system is to produce power. In the first case, this includes pressure reduction valve (PRV) replacement in industrial steam plant, refrigeration and air conditioning systems. In the latter case, heat from an external source is supplied to a system operating on a closed thermodynamic cycle and the working fluid is different to the heating and cooling media. A variety of applications and cycle configurations are described suitable for lower, medium and higher temperatures and where it is required to recover power from two heat sources within a single system, such as the jacket coolant and the exhaust gases of an internal combustion engine. The chapter concludes with a description of the factors that need to be taken into account in optimising system design.

Published by Woodhead Publishing Limited, 2014

Key words: screw expander lubrication, pressure reduction valve replacement, wet steam expansion, Organic Rankine Cycle systems, Wet steam cycle systems, system optimisation.

5.1 Screw expander lubrication

As indicated in Chapter 1, screw expanders have the advantage of being able to expand wet vapours. A description was also given of the oil-flooded and oil-free methods of compressor lubrication, and the fact that oil-flooded lubrication in compressors is the preferred method. This is both because it is cheaper and because it enables compressor efficiencies to be higher. The higher efficiencies are due to the smaller clearances possible with an oil film preventing direct rotor contact and consequent seizure and the cooling effect of the oil, which reduces the temperature rise associated with the compression process and hence makes it more nearly isothermal. It should be noted, however, that when expanding wet vapours there are serious drawbacks to oil-flooded lubrication. These are as follows:

1. When expanding wet vapours, it is practically impossible to separate the oil from the working fluid at the expander exit. It follows that in open systems, where the working fluid is not recirculated, there will be a continual loss of oil, while in closed cycle systems, the oil will be entrained with the working fluid and recirculated, thereby adversely affecting the heat exchanger effectiveness.
2. When expanding wet organic fluids, there is a strong tendency both for absorption of the oil by the working fluid and its dilution by the working fluid. This reduces the lubricant viscosity, and hence its effectiveness. Also,

when the oil concentration in the liquid component of the working fluid exceeds approximately 5%, as occurs during evaporation in a closed cycle system or during expansion, especially with higher oil to working fluid mass ratios, then the saturated vapour line of the working fluid becomes displaced. This implies that the evaporation temperature at a given pressure is increased. Evaporation is thereby inhibited and the performance of the expander system is thus reduced.

3. Oil-flooded lubrication, whether with wet or dry vapours, requires a recirculating pump to pressurise the oil leaving the expander, in order to reinject it at the high-pressure end of the expander. This is not required in a compressor, where the entrained oil is re-pressurised during the vapour compression process.

Most of these difficulties can be overcome by the use of our 'N' profile rotors, mainly due to the contact band between them being approximately at the pitch circle diameters of the rotors. Consequently, the relative motion between them is very nearly pure rolling, with very little sliding contact, while the contact forces between the rotors are low. As a result of this, it has been shown that the liquid content of the working fluid in two-phase expansion is sufficient to lubricate the rotors, even if the liquid viscosity is low. Thus, for wet vapour expansion, neither oil flooding nor timing gear is necessary.

The lubrication of the bearings may then still present problems. However, we have found that when the working fluid is organic and the expander operates within a closed cycle system, a small quantity of oil, of the order of 1% or less, by mass, dissolved in the working fluid can be sufficient to enable the bearings to be lubricated by the arrangement, shown in Fig. 5.1. Here, some of the cold pressurised fluid leaving the feed pump is directed directly to the bearing

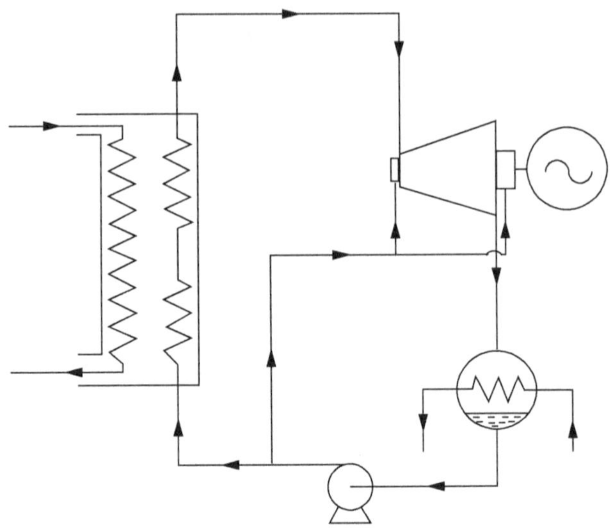

Figure 5.1 General layout of process fluid bearing lubrication for closed-cycle organic fluid power system

housing. There, frictional heating evaporates the working fluid and leaves sufficient oil in the bearings to lubricate them. At the same time, the very small percentage of oil in the working fluid has little adverse effect on either evaporation in the boiler or in reducing the power output of the expander.

5.2 Systems

Expanders can be used to generate power in any larger system that normally uses a pressure reduction valve (PRV). In most such applications, the general layout of the expander in the system is as shown in Fig. 5.2.

Although the principle of operation is similar in all cases, there are some differences, which should be noted for each application. Note that, in general, the expander is inserted in

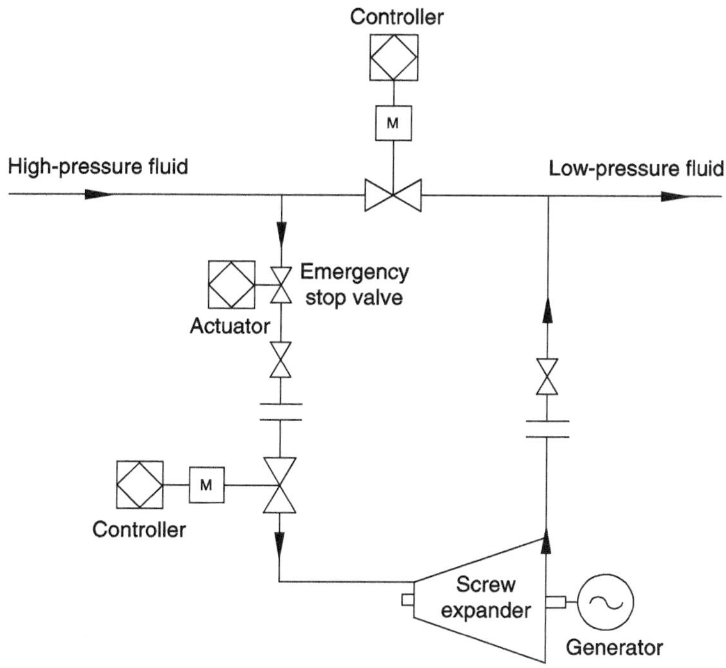

Figure 5.2 General layout of expander-generator within a larger system

parallel with the PRV. This is a safety requirement so that in the event of the expander failing for any reason, it is automatically shut down and the PRV activated, so that the main process, for which the pressure reduction is required, can continue uninterrupted. The following examples are cited to illustrate this.

5.2.1 Pressure reduction valve replacement in industrial steam processes

Many industrial processes require steam, including food industries and paper-making. It is common practice to raise the steam in a central boiler at relatively high pressure, typically in the range of 10–20 bar, to transport it to various sites in the factory and to reduce the pressure locally to a lower value, suitable for the process. Since raising steam at these higher pressures has a minor effect on the fuel consumption or the price of the boiler, considerable savings can be made in the total plant cost because the associated higher density steam can be transmitted in pipes of smaller diameter. The opportunity thus exists to replace the local PRVs by expanders and thereby recover power from the expansion process. The shaft power, thus extracted, can be used to drive a generator to produce electricity and thereby reduce electricity consumption from the public utility.

It is important to appreciate that, when the steam is required for a heating process, any power extracted from the steam must be compensated by the combustion of additional fuel in the boiler to increase the steam production rate so that the steam leaving the expander has the same energy as that leaving the PRV, from which no energy is extracted.

It should be noted that the efficiency of the process of conversion of heat to electricity, by this means, is approximately equal to the conversion efficiency of the mechanical power output from the expander to the electricity supplied to the grid, which approaches 90%, and is independent of the adiabatic efficiency of the expander itself. This is because the additional fuel burned only has to supply energy equal to the mechanical power extracted. Hence a low-efficiency expander will produce less power but will require less additional fuel to be burned to supply the required process heat. Nonetheless, a high expander efficiency for any given type of machine is desirable because more power will be produced for roughly the same size unit and hence the capital cost per unit output will be minimised.

The economics of this process then depend on the relative difference between the cost of electricity obtained from the utility and that of the fuel being burned. Typically, the cost per kilowatt-hour of electricity supplied by the grid is 3–4 times the cost per kilowatt-hour of heat derived from the burning of natural gas. Electricity generated by this means is, therefore, not obtained for zero fuel cost but for somewhere in the region of one-third that of electricity supplied by the utility.

Screw expanders have some distinct advantages for such applications, which are generally for power outputs of only a few hundred kilowatts, namely:

- they tend to be cheaper than turbines of equivalent power output;
- they have similar adiabatic efficiencies;
- unlike turbines, they can admit wet steam without loss in efficiency or reduction in service life.

An example of such a screw expander application is given in Fig. 5.3. As shown, the unit is totally enclosed and contains, within it, the expander coupled directly to a generator. It only requires connection to the steam supply and to the electricity mains. Start up and shut down are fully automated and, in the event of system failure, the steam supply to the expander is cut off and the PRV activated, thus making the whole industrial process fail-safe.

5.2.2 Pressure reduction valve replacement in flash steam geothermal power plant

Currently, worldwide, approximately 10,000 MW of electrical power is generated from steam, recovered directly from natural geothermal resources. In most cases, the steam, which may be in the form of pressurised water, below the surface, emerges at the well head as wet vapour, as a result of

Figure 5.3 A 100 kWe industrial process steam screw expander (courtesy of Heliex Power Ltd., East Kilbride, UK)

the pressure drop incurred as the fluid rises in the well. When the temperature of the wet steam emerging from the production well is above approximately 180 °C, the normal method of recovering power from it is to pass it through a PRV and then into a separator. The separated steam is passed through a turbine and the water is re-injected back into the ground through an injection well. Such a system is called a 'flash steam plant'. As shown in Figs. 5.4 and 5.5, the plant can be of the single or double flash type.

If the water–steam mixture could be expanded in a two-phase expander, the losses associated with the throttling of the steam and the reinjection of the hot separated water could be avoided and the efficiency of the process significantly

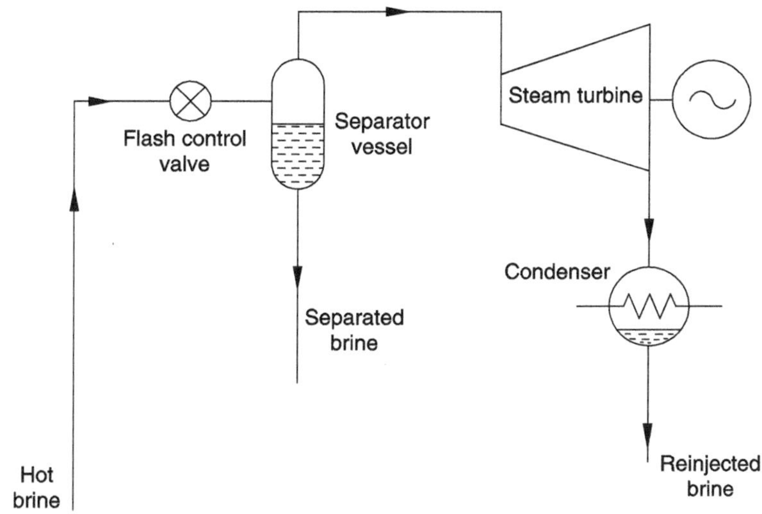

Figure 5.4 Single flash steam system for geothermal power generation

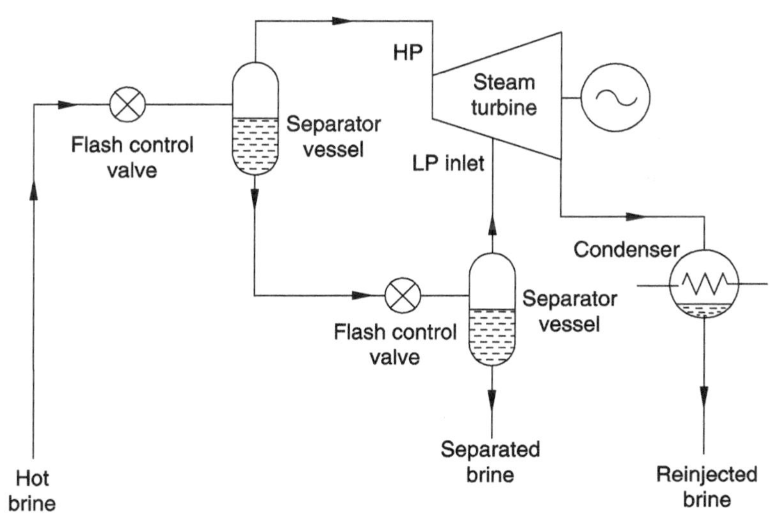

Figure 5.5 Double flash steam system for geothermal power generation

increased. This led to the 'total flow' concept in which, first screw expanders and then two-phase turbines of various types were investigated extensively in the 1970s and early 1980s. A variety of problems were encountered with both types of expander and this approach has been largely abandoned. However, the possibility of replacing the PRVs by screw expanders in existing flash steam plant still exists, as shown for a single flash system in Fig. 5.6.

In a dual flash system, it is almost certain that it would only be economically viable to replace the high-pressure PRV because the relatively small pressure drops and large volumes associated with the second throttling process would result in an inefficient and excessively large screw expander. The arrangement for this is shown in Fig. 5.7.

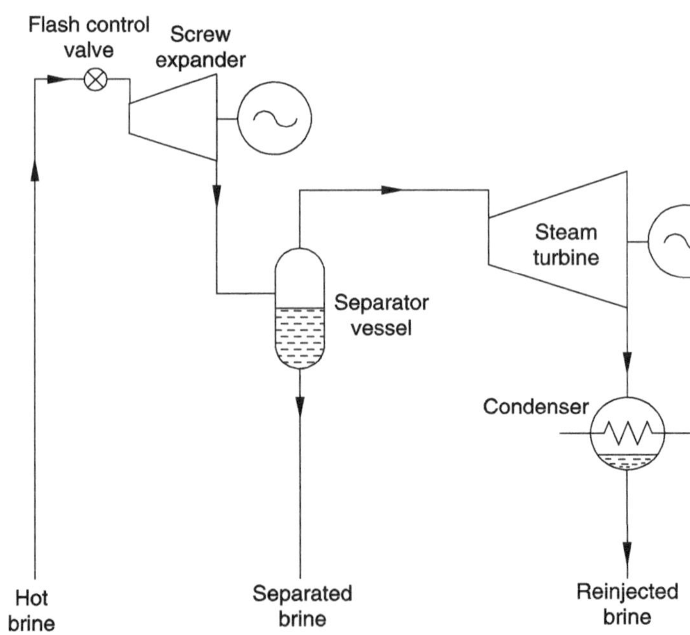

Figure 5.6 **Single flash steam system with screw expander**

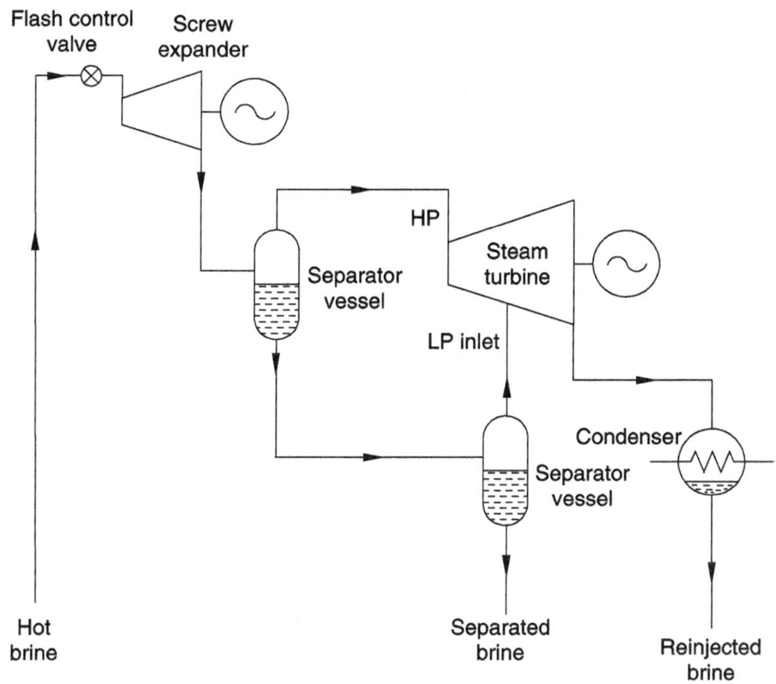

Figure 5.7 **Double flash steam system with single screw expander**

There are, however, problems associated with PRV replacement in such systems. First, the brines, emerging from the ground, usually contain significant amounts of silicates and salts, which tend to come out of solution as the pressure and temperature are reduced. Second, the relatively low pressures at which these systems operate and the associated large specific volume of steam result in the need for very large, and hence expensive, machines.

Proposals for such an application have frequently been made but, at the time of writing, no installations are known to the authors.

5.2.3 Pressure reduction valve replacement in natural and process gas transport

Natural gas, which is mainly methane, is transported to its users through pipes at high pressure, in order to minimise the cost of the pipe network, by minimising the pipe size needed. At certain points, it is depressurised for local distribution. As in the case of process steam, power can be recovered from the depressurisation by replacing the PRVs used for this purpose by screw expanders or by turbines for higher outputs. The layout for such a system is shown in Fig. 5.2. However, in most cases, there would need to be an important additional component. This is due to the fact that the expansion is associated with a large temperature drop. Natural gases are almost never pure and often contain traces of water. Starting from approximately atmospheric temperature, expansion could therefore lead to components freezing during the expansion process. To avoid any problems associated with this, the gas needs to be preheated, which can be done by burning some of the gas transmitted. Care has then to be taken to ensure that all equipment is fully flame-proofed to meet the requirements of the hydrocarbon processing industries.

In some cases, especially in small offshore oil rigs, gas expanders of this type can be used to generate an uninterrupted power supply (UPS) for the oil rig electrical power requirements. In this case, the frequency of the electricity generated cannot be controlled using an induction generator, connected to the main grid, as is possible in industrial waste heat applications, where the power generated is generally small compared with the total power requirement. More elaborate controls are then needed for the expander–generator set to ensure that the power generated is supplied at a constant frequency.

5.2.4 Pressure reduction valve replacement in refrigeration and air conditioning systems

Vapour compression systems are used for refrigeration and air conditioning. The main components are shown in Fig. 5.8. As can be seen, the liquid leaving the condenser is cooled by passing through the throttle valve, where the pressure drops, resulting in partial vaporisation of the refrigerant and a significant reduction in temperature. If the throttle valve is replaced by an expander, then some power can be recovered and the refrigeration effect of the fluid passing through the evaporator will be increased because less vapour will be formed if work is extracted during the two-phase expansion process, thus giving a twofold advantage.

In domestic refrigeration units, the amount of power that can be recovered is far too small to justify replacing the throttle valve by an expander.

In larger units, driven by scroll or screw compressors, improvements to the coefficient of performance (CoP) can be made by expanding the vapour in two stages, separating the

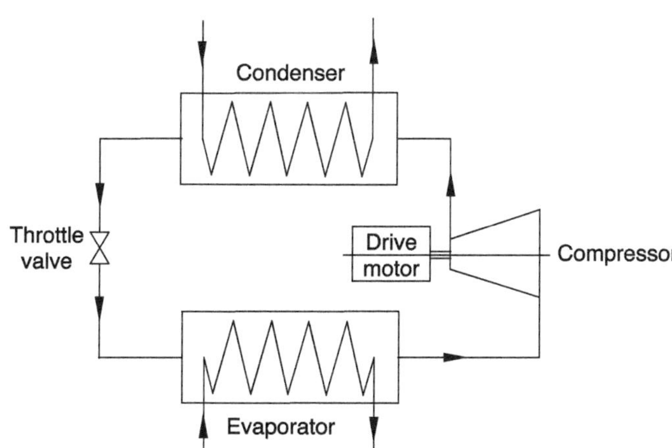

Figure 5.8 Basic vapour compression refrigeration system

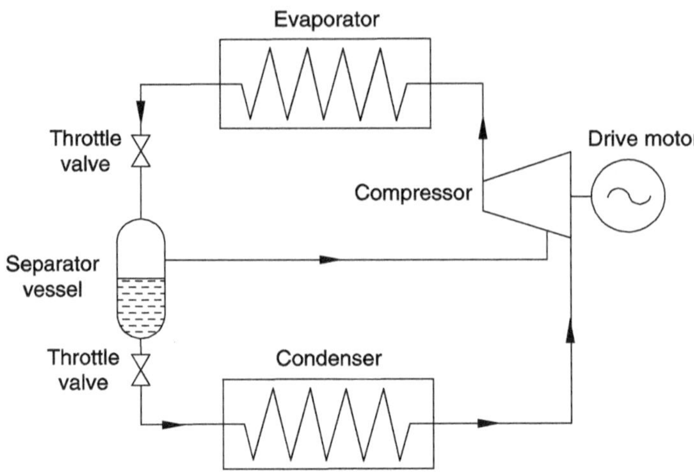

Figure 5.9 **Vapour compression refrigeration system with economiser**

vapour from the liquid at the intermediate pressure and admitting the separated vapour into the compressor at a convenient position, after the suction port has closed and compression started, as shown in Fig. 5.9. Such an arrangement is called an 'economiser'.

Large-scale air conditioning units often use centrifugal compressors for the compression process. In that case, it is not possible to admit vapour into the compressor at an intermediate pressure and hence an economiser cannot be employed. Under these circumstances, where the recoverable power from the two-phase expansion process can be significant, it is worth considering means of doing this. We have demonstrated one means of achieving this, using a device that we have called an 'expressor'.

The basic concept is that the power recovered by replacing the throttle valve by a two-phase expander is used to recompress part of the vapour formed in the two-phase expansion process by directly coupling it to a screw compressor so that the unit runs independently of the main

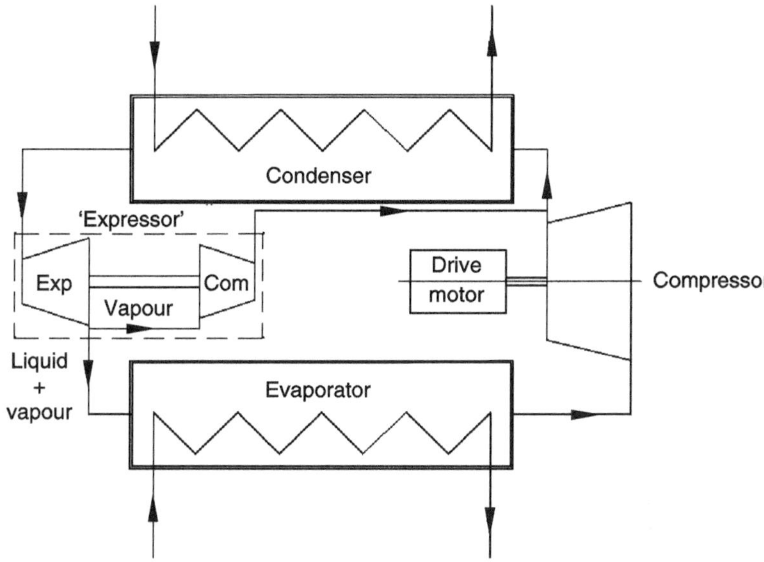

Figure 5.10 An expressor in a vapour compression system

compressor, thereby eliminating the restraints created by coupling to the main compressor or the losses associated with converting the power generated to electricity if the expander were to drive a generator. This arrangement is shown in Fig. 5.10.

The concept can be further improved if the expansion and compression processes are both carried out on the same pair of rotors. Either of two methods can be used for this purpose, the first of which is based on the fact that increasing the wrap angle of a twin-screw machine does not increase the maximum trapped volume but simply extends the rotational angle at which the maximum volume is maintained and hence increases the time for which the working chamber of the expander is exposed to the low-pressure port.

This principle, shown in Fig. 5.11, can be used to allow expansion to start at one end of the rotors and to be

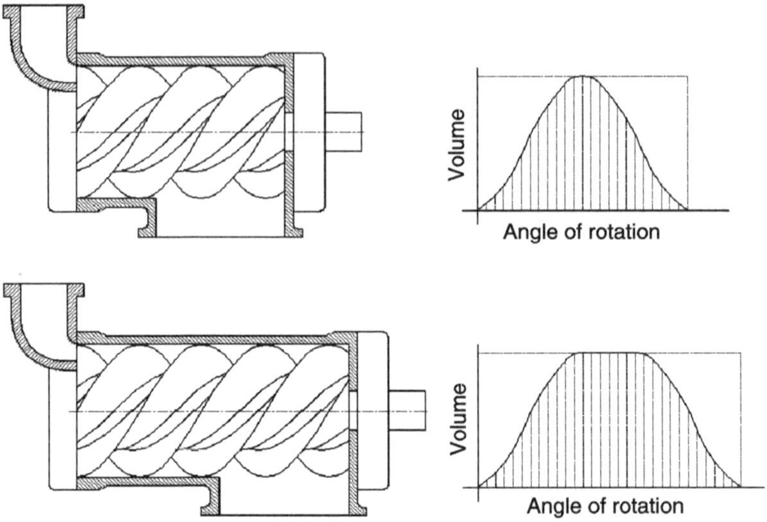

Figure 5.11 The effect of the wrap angle on the trapped volume

completed part way along their length. At this point, the rotors are exposed to an intermediate low-pressure port in the casing and exposed to it for a sufficient time for the liquid and excess vapour to be expelled from the casing and enter the evaporator. Further rotation then traps the remaining vapour within the working chamber and results in this vapour being recompressed and delivered to the condenser on leaving the second high-pressure port at the other end of the rotors (see Fig. 5.12). The resulting form of the expressor is shown in Fig. 5.13. The rotors and casing for this are shown in Fig. 5.14.

Assembled views of two early units are shown in Fig. 5.15. The unit in the upper panel carried over too much liquid into the compression section. This was overcome by enlarging the low-pressure port, as shown in the lower panel of the figure.

The second approach to the expressor concept is to perform the same expansion and compression processes

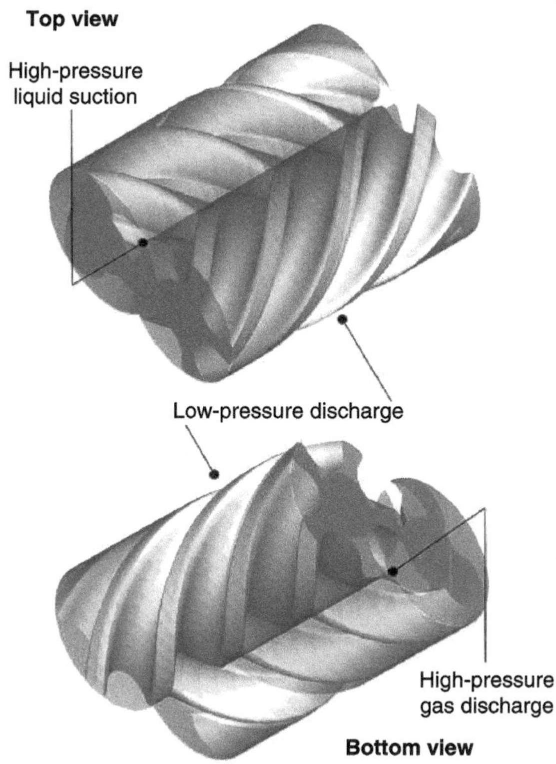

Figure 5.12 Expansion and recompression in one pair of rotors

with a single rotor pair but, in this case, to separate the two functions by a partition so that there is no carryover of liquid from the expansion section into the compression section. A view of the rotors required for this is shown in Fig. 5.16.

It should be noted that the rotors required for an expressor require a different profile to those used for compression or expansion processes only. This is because both sides of the rotors are required to have a line of contact. The resulting profile is, therefore, nearly symmetrical, as shown in Fig. 5.17.

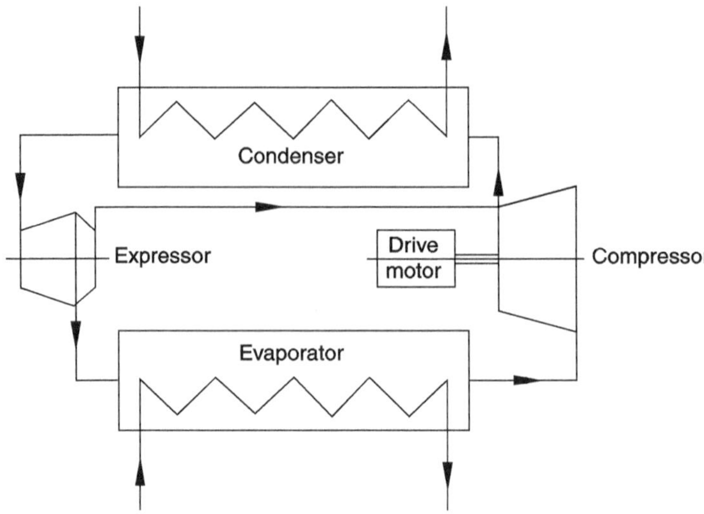

Figure 5.13 The expressor as a single rotor pair unit

Figure 5.14 Expressor components

Figure 5.15 Prototype expressor units

Figure 5.16 Expressor rotor pair for separating the working chamber into two sections

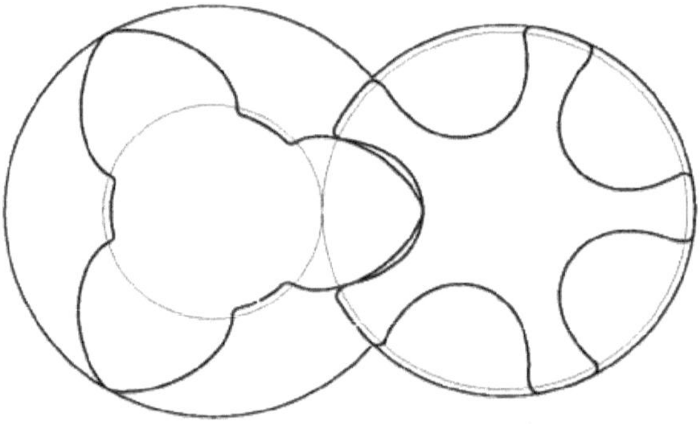

Figure 5.17 Expressor rotor profile

A cross-section of an expressor unit with separate expansion and compression working chambers is shown in Fig. 5.18. The partition is made in two halves, which are fitted round the rotors and which slot into each other to ensure a seal between them.

A separator plate in a prototype expressor casing, with separate working chambers, is shown in Fig. 5.19. This was designed to replace the throttle valve in a 500 ton centrifugal chiller unit. Test rig performance results indicated that the

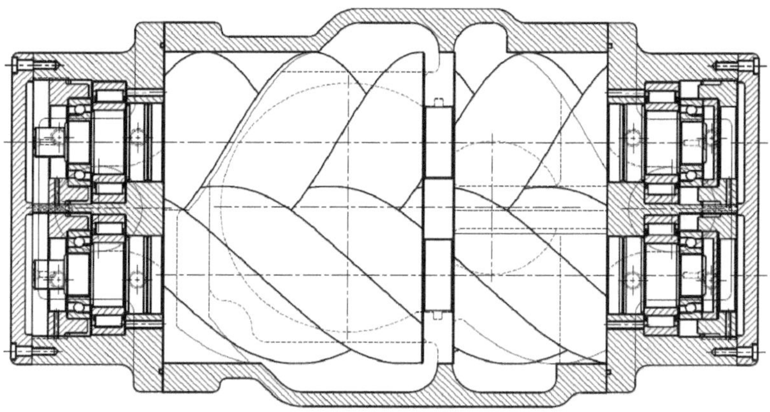

Figure 5.18 Cross-section of an expressor with separate expansion and compression working chambers

Figure 5.19 Expressor casing with separate expansion and compression working chambers (view from gas discharge end)

use of this prototype in place of a throttle valve would improve the coefficient of performance of the chiller unit by 8–10%.

5.2.5 The compressor-expander

A fairly obvious extension of the expressor concept is possible, in systems where power is required to drive a screw compressor, some of which can be recovered in the expansion section. In principle, this could be applied to any vapour compression system driven by a screw compressor, as shown in Fig. 5.20. This can confer some advantages, in that the

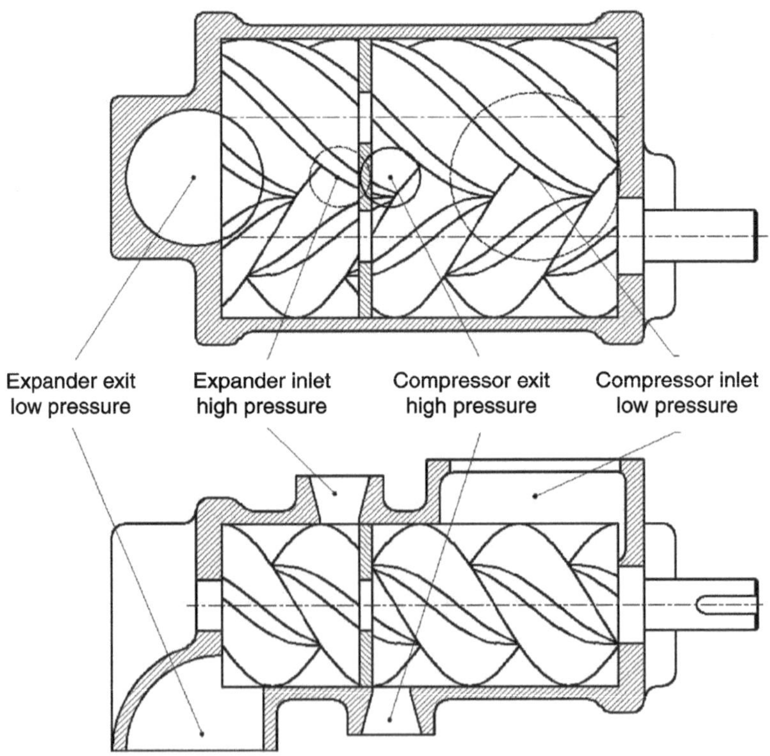

Figure 5.20 Screw compressor-expander

large axial loads on the bearings can be greatly diminished, while the radial bearing loads can be significantly reduced.

In practice, when halogenated hydrocarbon working fluids or their equivalent are used, the ratio between the compression work and that recoverable from the two-phase expansion process is too large. This means that the axial length of the rotors dedicated to expansion is too short compared with that required for compression. Tests on such an arrangement showed that, as a result, the leakage rates in the expansion side are too high and no significant performance benefits could therefore be obtained from such an arrangement. However, there are possibilities for such a unit in some cases.

One such case is for a vapour compression cycle system using carbon dioxide as the working fluid. Due to the very high pressure differences at which this is required to operate, the throttling losses are very large and the resulting coefficient of performance is very low. In this case, the ratio between the compression work and the recoverable expansion work is more favourable. The arrangement required is shown in Fig. 5.21. Note how the low- and high-pressure ports are arranged to balance the bearing loads as far as possible and to minimise leakage across the separating plate inside the casing.

The large losses incurred as a result of throttling in an ideal transcritical CO_2 cycle system are shown in Fig. 5.22, and estimates of what improvement can be made in a practical subcritical cycle, using a combined compressor expander, in Fig. 5.23. As can be seen, improvements in the coefficient of performance of the order of 16% are possible by such means.

Another potential application of this concept is to enhance the performance of fuel cell systems. In this case, the pressure in the fuel cell has to be raised above atmospheric pressure in order to increase the voltage and keep the cell compact.

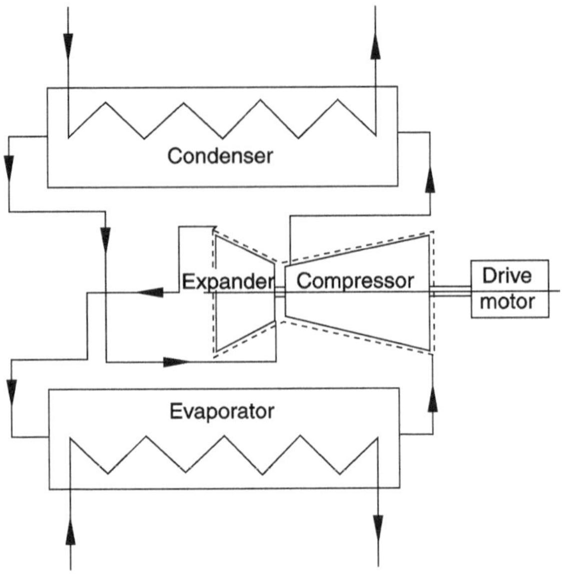

Figure 5.21 Carbon dioxide refrigeration system

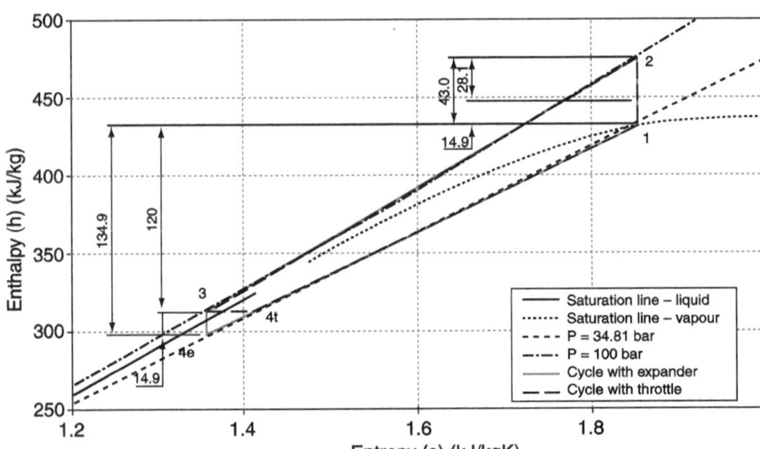

Figure 5.22 Estimated performance improvements in an ideal transcritical CO_2 cycle system with combined compression and two-phase expansion. $\Delta Q_{evap} = 11\%$, $\Delta W_{in} = -35\%$, $\Delta CoP = 72\%$

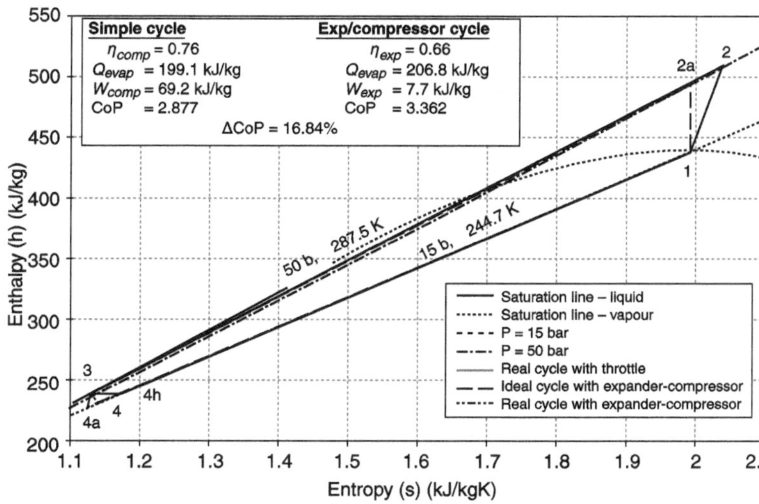

Figure 5.23 Estimated performance improvement in a subcritical CO_2 system using a compressor expander

Typically, the cell would operate at 3 bar pressure and the air compressor required for this absorbs approximately 20% of the cell output.

Considerable savings could be made by expanding the cell reaction products, mainly nitrogen and steam, to reduce the required power input in a compressor-expander. A compressor-expander unit, designed for this purpose, is shown in its original form in Fig. 5.24. Problems were encountered with this unit, due to differential expansion between the rotor and the casing causing rotor contact with the plate separating the compressor and expander sections. These were finally overcome by constructing the compressor section and expander section casings of two different materials, with the compressor casing made from grey cast iron and the expander casing from aluminium, as shown in Fig. 5.25.

Figure 5.24 Prototype fuel cell compressor-expander components

Figure 5.25 Fuel cell compressor-expander revised casing design

5.3 Cycles

Most systems for power recovery from low-grade heat, whether in use or proposed, are based on the Rankine cycle or sub-variants of it. As will be shown, a number of these are made possible due to the ability of screw expanders to expand liquid–vapour mixtures with efficiencies approaching those of turbines.

It should be borne in mind that because fluid velocities in screw expanders are lower than those in turbines, they tend to be larger for a given power output. Also, the manufacturing processes currently available restrict the maximum rotor diameter with acceptable tolerances to 800–900 mm. Overall, this limits the maximum power output that can be obtained from such machines to about 5 MW. Unfortunately, although it is possible to produce such large rotors, their manufacturing cost rises rapidly at diameters much above 350 mm. At lower power outputs, where smaller machines are required, the internal clearances imposed by manufacturing limitations become relatively large compared with the machine dimensions. Consequently, flow through internal leakage becomes an increasing part of the total, and expander efficiencies become unacceptably low. Thus, the preferred operating range of power outputs for screw-driven power plant is approximately 50 kW to 1 MW.

As already shown in Chapter 1, especially at lower source temperatures, organic working fluids convey a number of advantages including improved matching between the heating medium and working fluid temperatures and improved expander efficiencies due to dry vapour expansion, throughout, when using turbines. The disadvantage of such fluids is that they are usually superheated at the expander exit. As shown in Chapter 1, the desuperheat can be usefully employed by recovering it in a recuperator to preheat the working fluid leaving the feed pump. However, in extreme cases, when operating with high maximum temperatures, the temperature of the pressurised liquid leaving the recuperator may be so high that it reduces the heat recoverable from the heating medium.

The choice of working fluid is dependent both on the power output and the temperature range over which the power plant operates, generally with higher critical temperature fluids being required for hotter heat sources. Additionally, environmental concerns restrict the choice of working fluid. Consequently, there is a continuing search for fluids that combine good thermodynamic properties with low environmental impact.

5.3.1 Organic working fluid cycles

Saturated Rankine cycle

This is the basic Organic Rankine Cycle (ORC), as used in most ORC systems. It consists of four main components only, as illustrated in Fig. 5.26. When working over moderate temperature differences, as in most geothermal applications, the degree of superheat of the fluid leaving the expander is not sufficient to justify a recuperator, especially since its inclusion would reduce the expander power output, due to the raised back pressure created by flow losses within the recuperator. As already explained, the use of organic fluids enables the turbine efficiency to be increased, due to the absence of liquid in the expansion process, the lower specific energy of high molecular weight fluids enables the expansion to be performed in fewer turbine stages, normally not more than one, and the higher condensing pressure, due to the lower critical temperature of such fluids, results in the expanded vapour being more dense and thus allows the expander to be smaller.

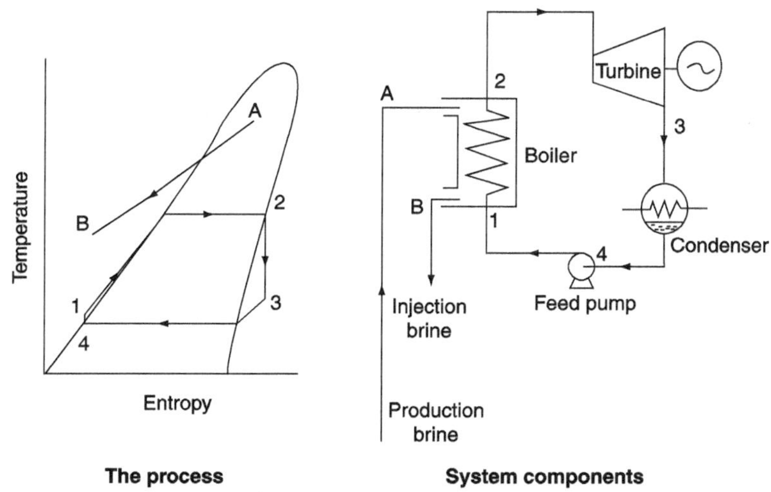

Figure 5.26 **Saturated Rankine cycle system with organic working fluid**

It should be noted that as the source temperatures increase, fluids with higher critical temperatures are required to enable part of the heat to be transferred in the evaporative mode. This must be associated with lower condensing pressures and hence, apart from thermal stability and flammability considerations, the advantages of smaller expanders and fewer stages are progressively reduced.

Wet Rankine cycle

In general, at the lower end of resource temperatures, there is insufficient desuperheat at the end of expansion to justify the use of a recuperator in a saturated Rankine cycle system. Thus, by using a screw expander, instead of a turbine, the output can be increased and the overall system conversion efficiency improved by admitting the working fluid to the expander in the slightly wet condition, so that it leaves the expander as dry saturated vapour, as shown in Fig. 5.27.

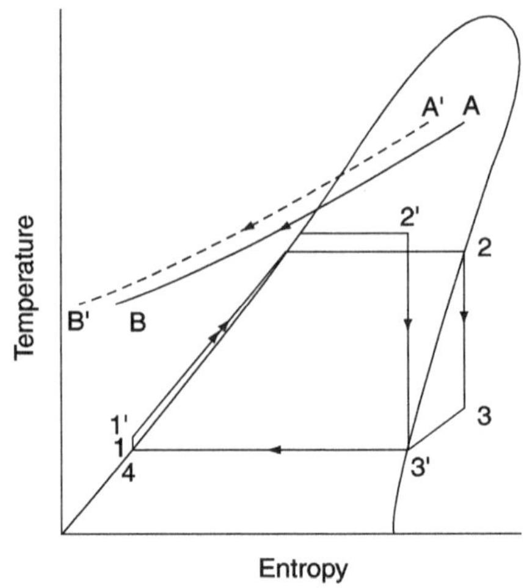

Figure 5.27 Wet organic Rankine cycle (WORC)

Benefits accrue from higher cycle efficiency, due to the increased evaporation temperature and greater mass flow rate, due to less evaporative heating. The disadvantage is that the specific enthalpy drop of expanding wet vapour is less than that of dry vapour. Overall, in most cases, gains in output of the order of 5% are possible with no cost incurred. A WORC system, currently manufactured and operating successfully in a number of installations, is shown in Fig. 5.28.

Trilateral flash cycle

The trilateral flash cycle (TFC) is shown in Fig. 5.29. Its advantages and disadvantages have been reviewed in Chapter 1. Although it maximises the recoverable power, its main disadvantage is that the heat transferred per unit output is relatively large, due to the poor conversion efficiency of the additional heat recovered at the lower temperatures. This could be justified in applications such as enhanced geothermal systems (EGS), i.e. man-made geothermal resources created

Figure 5.28 A 50 kWe screw expander-driven industrial WORC system (courtesy of Electratherm Inc., Reno, Nevada, USA)

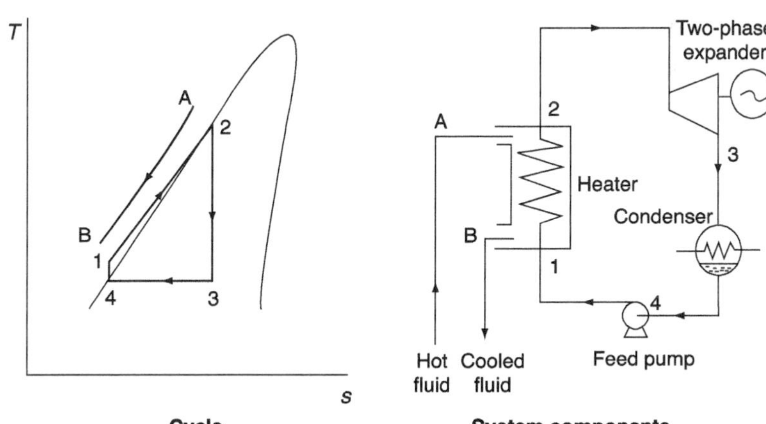

Figure 5.29 Trilateral flash cycle (TFC)

by drilling and fracturing rock to create an underground region of cracks, through which water could be circulated from the surface to recover heat from the rock. In such cases, the high cost of creating the resource could justify additional expenditure on the surface plant to maximise the power recovery.

The difficulty arises at higher resource temperatures, where the volume ratio of expansion of the working fluid is too large to be accommodated in a single screw expander stage. The second stage expander may then need to be too large to be cost-effective or beyond current manufacturing limits.

Higher temperature two-phase (Smith) cycle

The cycle shown in Fig. 5.30 was proposed by the authors to overcome the problems associated with the need for two expander stages for the TFC at higher resource temperatures in the 180–210 °C range. Studies with pentane as the working fluid showed that a single screw expander stage is sufficient for the two-phase expansion, while a conventional turbine can be used for the dry vapour expansion. Preheating of the working fluid leaving the first stage feed pump by the liquid separated from the screw expander exhaust raises the temperature at which it enters the primary heat exchanger. This reduces the recoverable heat but improves the cycle efficiency. Analyses have shown that such a system can attain second law efficiencies of nearly 50%, which indicates that it could be an attractive system for EGS applications, where brines can be obtained at these temperatures.

At higher temperatures, there can be considerable desuperheat needed at the end of the turbine expansion process. In such cases, a recuperator can be included as shown in Fig. 5.31. At these temperatures, the cycle efficiency of this arrangement is comparable to that of saturated,

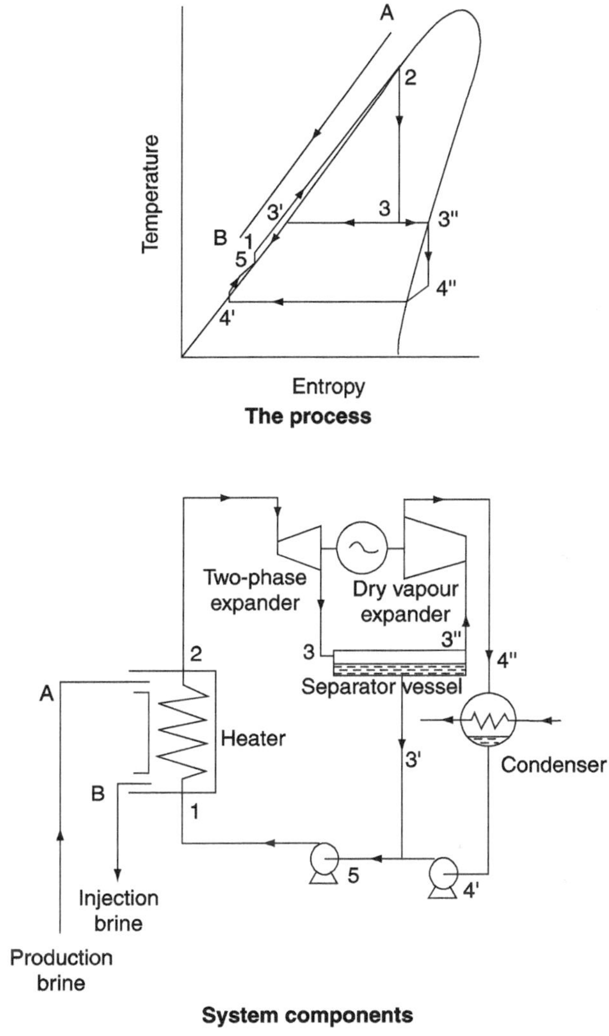

Figure 5.30 Higher temperature two-phase expansion cycle system

superheated or supercritical ORC systems, and hence the condenser cooling requirements are of the same order, while comparative studies show that the recoverable power appears to be 5–10% greater. Size limitations for the screw expander

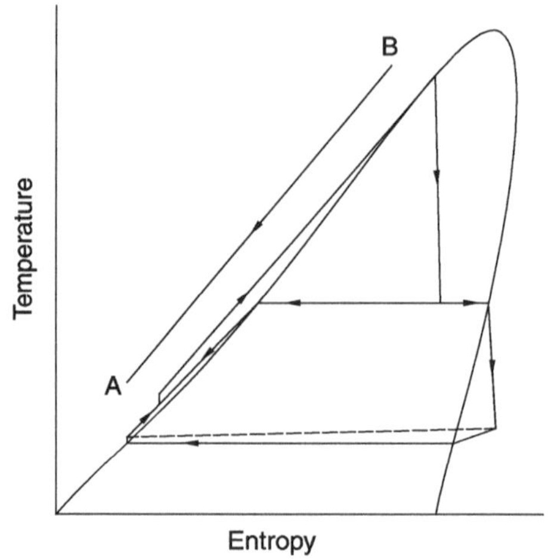

The cycle

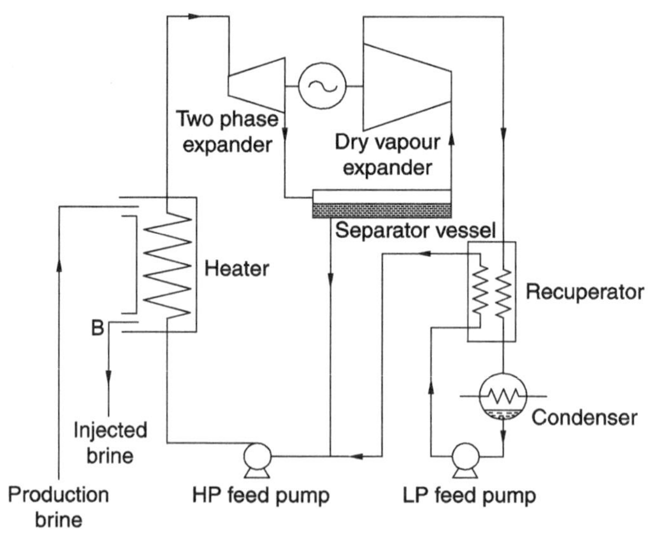

System components

Figure 5.31 Recuperated higher temperature two-phase expansion system

indicate that the upper limit for a single screw–turbine combination of this kind is of the order of 10 MW.

Supercritical cycle

This has already been described in Chapter 1 and is only included here for some further comment. As shown in Fig. 5.32, recuperated supercritical cycle systems have been used successfully for power recovery from medium enthalpy geothermal resources using R134a as a working fluid. Such systems offer a number of advantages in that working fluids with lower critical temperatures are used and this implies that condensing pressures will be higher and pressure ratios, across the expander, lower. Accordingly, the working fluid will be denser, and hence the components will be more compact, while fewer turbine stages are required. Additionally, operating at supercritical pressures gives good temperature matching between the brine and the working fluid in the primary heat exchanger.

Most organic working fluids have critical pressures in the 30–40 bar range. This implies that maximum pressures in the cycle will be above 40 bar and the feed pump work will be relatively large. With these requirements, there are a number of issues that make screw expanders unsuitable for use in such systems. First, there are problems with designing bearings that can withstand the large pressure differences required in the expanders. Second, the screw expander efficiencies are lower than those of turbines. Third, on the assumption that an all-screw expander system will not exceed 1 MW in power output, the working fluid flow rates would be relatively small and under conditions of low flow rate and high pressure rise, feed pumps are less efficient. Accordingly, with relatively higher feed pump work and less expansion work, the system work ratio is rather low and this

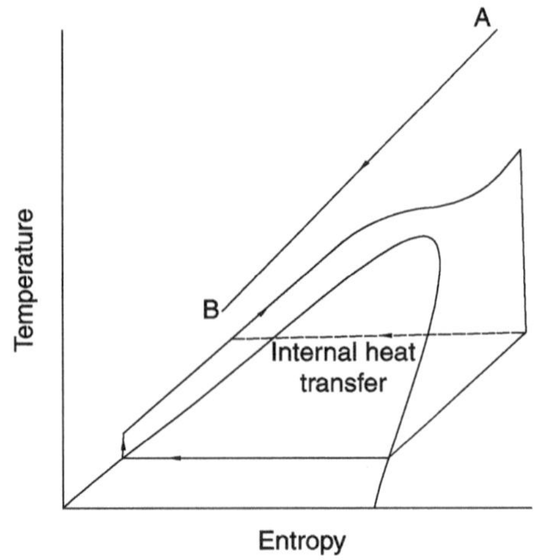

The cycle

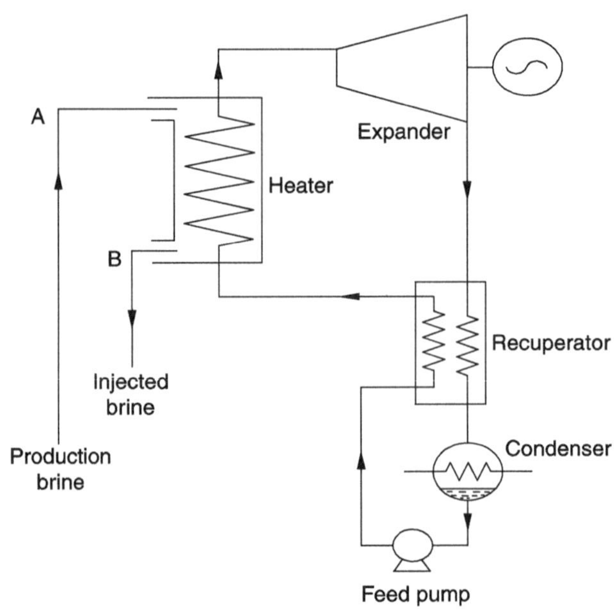

System components

Figure 5.32 Recuperated supercritical ORC system

implies that the overall conversion efficiency will usually be lower than that of an alternative subcritical cycle system.

Supercritical carbon dioxide cycle systems have also been proposed, operating at very high pressures and are being developed at the time of writing. We are unclear how these will function effectively as Rankine cycle systems in hot weather, since the critical temperature of CO_2 is 31 °C. It follows that condensation will not be possible with coolant temperatures above approximately 20 °C. This precludes their operation as air-cooled systems in hot climates and at normal summer conditions in most countries, or in marine applications in tropical seas.

Superheated cycle

As shown in Fig. 5.33, superheated ORC cycles almost always require a recuperator to desuperheat the working fluid after expansion.

Although the advantages of a recuperator have been already cited, a further feature of it, which has also been noted, is that there is inevitably a pressure loss in the fluid passing through it and this raises the exit pressure of the expander, thus leading to some loss of power output and cycle efficiency. This loss is generally small compared with the overall gain in the use of the recovered heat but it should not be overlooked in evaluating such systems.

Superheated ORC systems are not widely used because overall heat transfer coefficients across the recuperator between the working fluid in the low-pressure vapour and high-pressure liquid phases are low, while the log mean temperature difference between the phases is small. Recuperators are therefore relatively large and expensive. Those more favoured are saturated vapour systems using working fluids with higher critical temperatures.

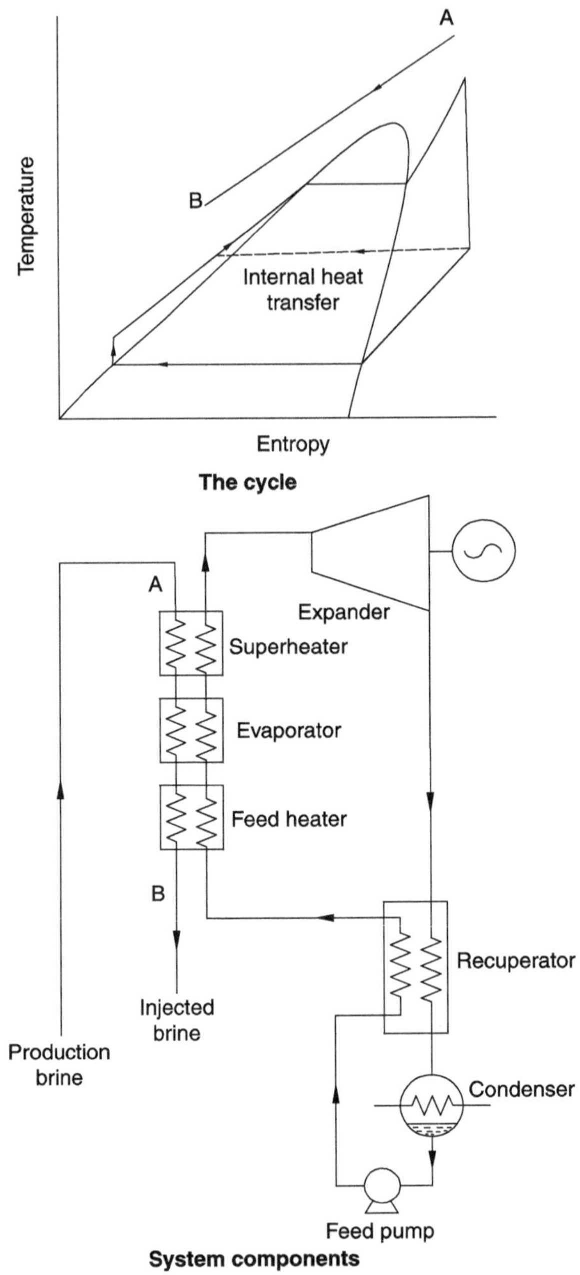

Figure 5.33 Superheated ORC system with recuperator

Regenerative feed heating

Regenerative feed heating, where steam is drawn off the turbine between stages to preheat the feed water, is a well-established practice in steam power plant. The two methods employed are shown in Figs. 5.34 and 5.35. However, its advantages and disadvantages are rarely clearly explained.

In large-scale power plant, the maximum operating pressure and temperature are set at values determined largely by metallurgical limits and not by thermodynamic considerations. Accordingly, when steam is bled off at intermediate stages of the turbine, its function of preheating the feed water does not affect the pressure and temperature limits over which the plant operates. It raises the cycle efficiency by reducing the external heat supplied by the boiler but at the expense of reducing the power output. Thus, unless some other use is made of the available heat from the flue gases, which are then not required to preheat the feed water, regenerative feed heating results in a reduction in the plant power output and hence overall efficiency. However, in the case of steam power plant, the now available but unused heat from the flue gases can be used to preheat the air entering the boiler, thereby reducing the fuel requirement, and hence in accordance with Carnot's principle, raising the overall plant thermal efficiency far more effectively than if it were used to preheat the boiler feed water. Thus, without the available alternative use for the low-temperature boiler exhaust gases, the use of regenerative feed heating in large steam power plant would not be advantageous.

In the case of low-temperature systems, where the cycle temperature and pressure limits are not restricted by metallurgical considerations, regenerative feed heating can be used advantageously, as is the case with recuperative heating, provided that it does not result in a reduction in the recovery of heat from the high-temperature source.

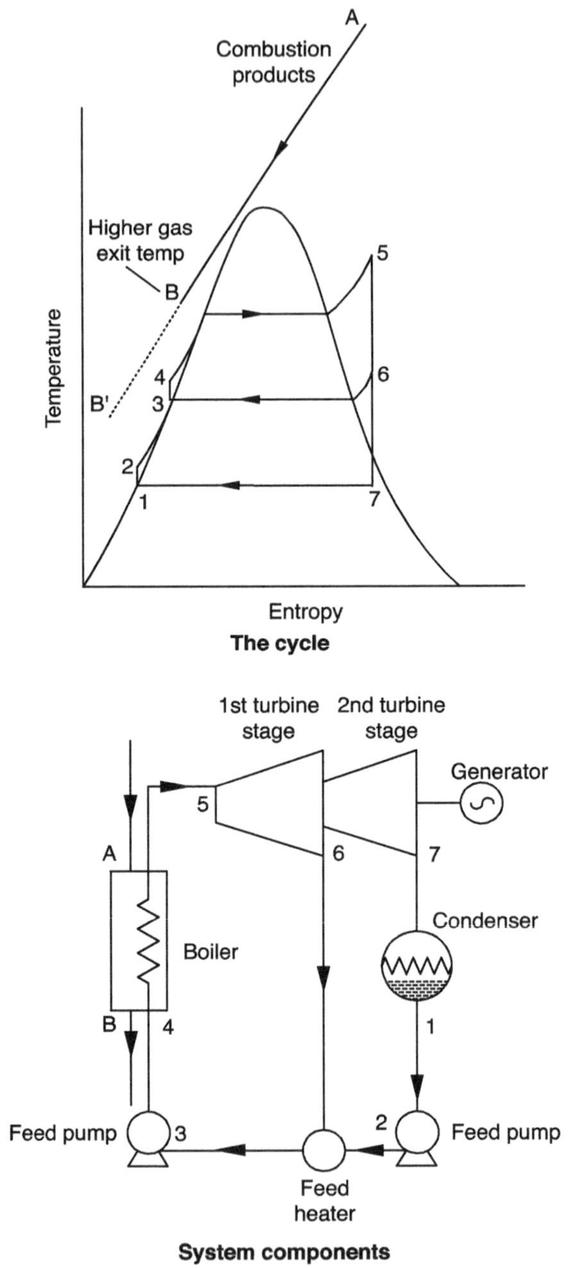

Figure 5.34 Direct contact regenerative feed heating

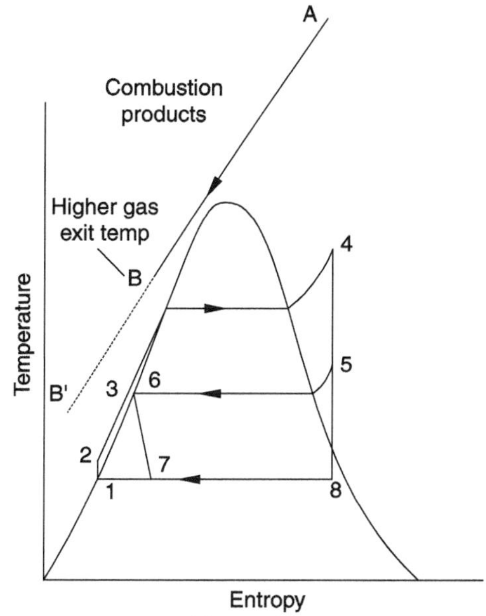

The cycle

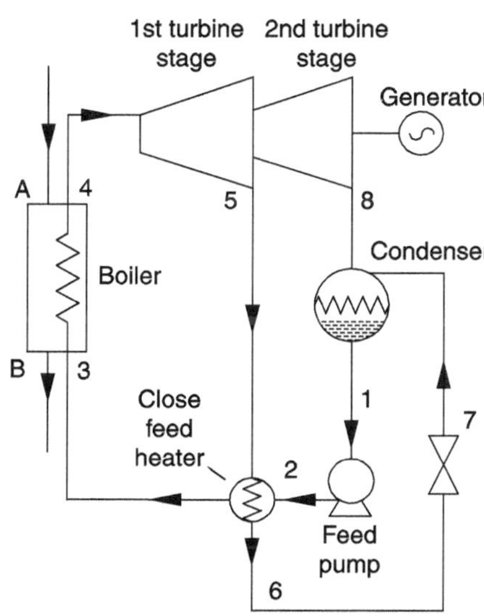

System components

Figure 5.35 Indirect contact regenerative feed heating

In turbine-driven plant, where the expansion is carried out in a single radial inflow stage, it is not possible to bleed off the expanding vapour at an intermediate pressure. However, when the expansion is carried out in two or more stages, or where a screw or other type of positive displacement expander is used, it is possible to include regenerative feed heating, as shown in Fig. 5.36.

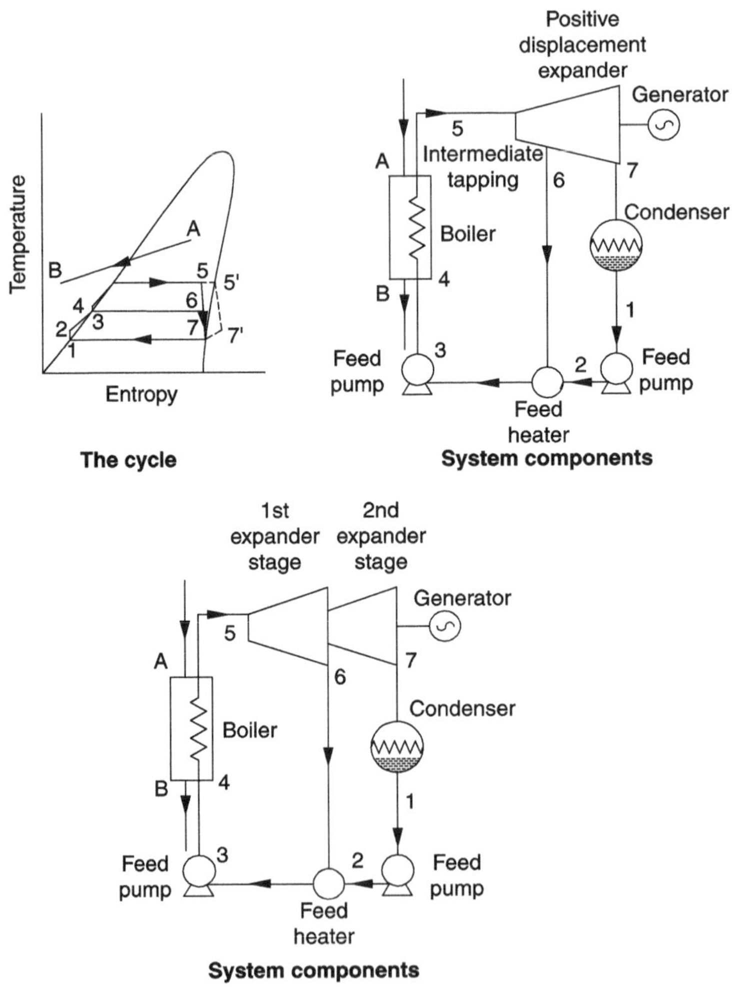

Figure 5.36 (W)ORC system with regenerative feed heating obtaining heat from a low-grade heat source

Such an arrangement can be particularly advantageous where the temperature of the recovered heat is restricted to a value well above the condensing temperature. Examples of this are geothermal brines, where dissolved solids must not be allowed to precipitate, in recovering power from IC engine cooling systems, where there are defined temperature limits for the coolant, set by the engine manufacturer, or from IC engine exhaust gases, where the exhaust gases may contain sulphur or other impurities. In such cases, cycle efficiency and power output improvements of the order of 5–10% are possible.

Cycles with multiple heat sources

The main function of such cycles is to maximise power recovery from IC engine waste heat, utilising the heat recoverable from both the exhaust gases and the cooling jacket and, where applicable, the supercharger and oil coolers.

Dual pressure cycle

For this purpose, it is possible to adapt the dual pressure cycle, as used in the steam section of large combined cycle gas turbine plant (Fig. 5.37). This has the advantage of using the same working fluid for both the lower and higher temperature heat recovery and thereby minimising the number of heat exchangers required. By making use of the screw expanders' ability to expand wet vapour, the system can be improved to eliminate the need to desuperheat the working fluid at the end of either expansion stage, as shown in Fig. 5.37. The overall system arrangement for this is shown in Fig. 5.38.

There are two disadvantages to this system. The first is that the need to use a working fluid with a relatively high

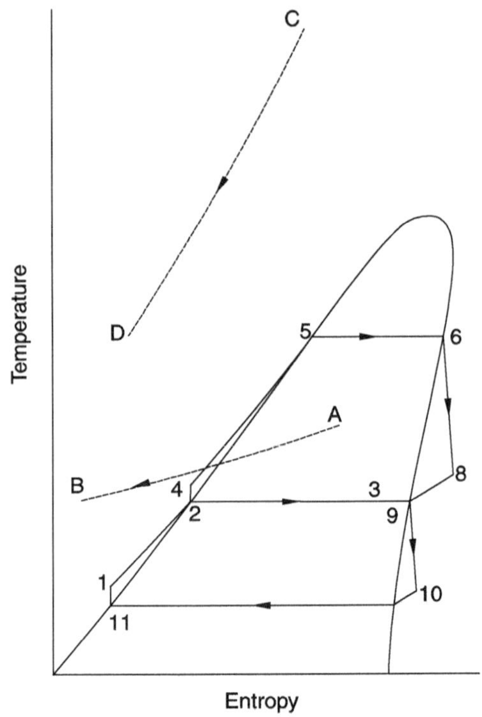

The cycle

Figure 5.37 Dual pressure cycle with dry vapour expansion

critical temperature, in order to recover the high-temperature heat efficiently, leads to the need for a much larger and more expensive expander in the lower pressure stage. The second disadvantage is less obvious but more significant: in most IC engines, the heat rejected to the cooling jacket and intercooler predominates and this limits the optimum performance of the higher pressure section in order to obtain a thermal balance at the exit of the high-pressure expander when the expanded high-pressure fluid and the unexpanded intermediate pressure fluid combine. Overall, the efficiency is therefore rather less than if two separate systems were used,

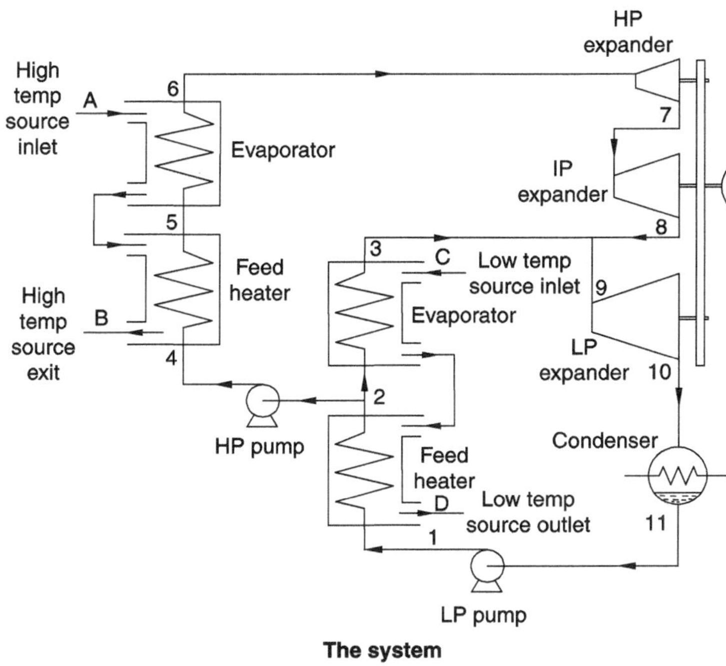

Figure 5.38 Dual pressure cycle system

one for the high-temperature heat recovery and the other for the lower-temperature heat recovery.

Binary fluid system

An alternative to the dual pressure cycle is to superimpose two Rankine cycle systems with separate topping and bottoming cycles, using the condenser of the topping cycle as the evaporator of the bottoming cycle, as shown in Fig. 5.39. This involves an additional heat exchanger and two separate feed pumps, instead of the two-stage feed pump of the dual pressure cycle. It should be noted, however, that the intermediate heat exchanger, acting as the topping cycle condenser and the bottoming cycle evaporator, will be

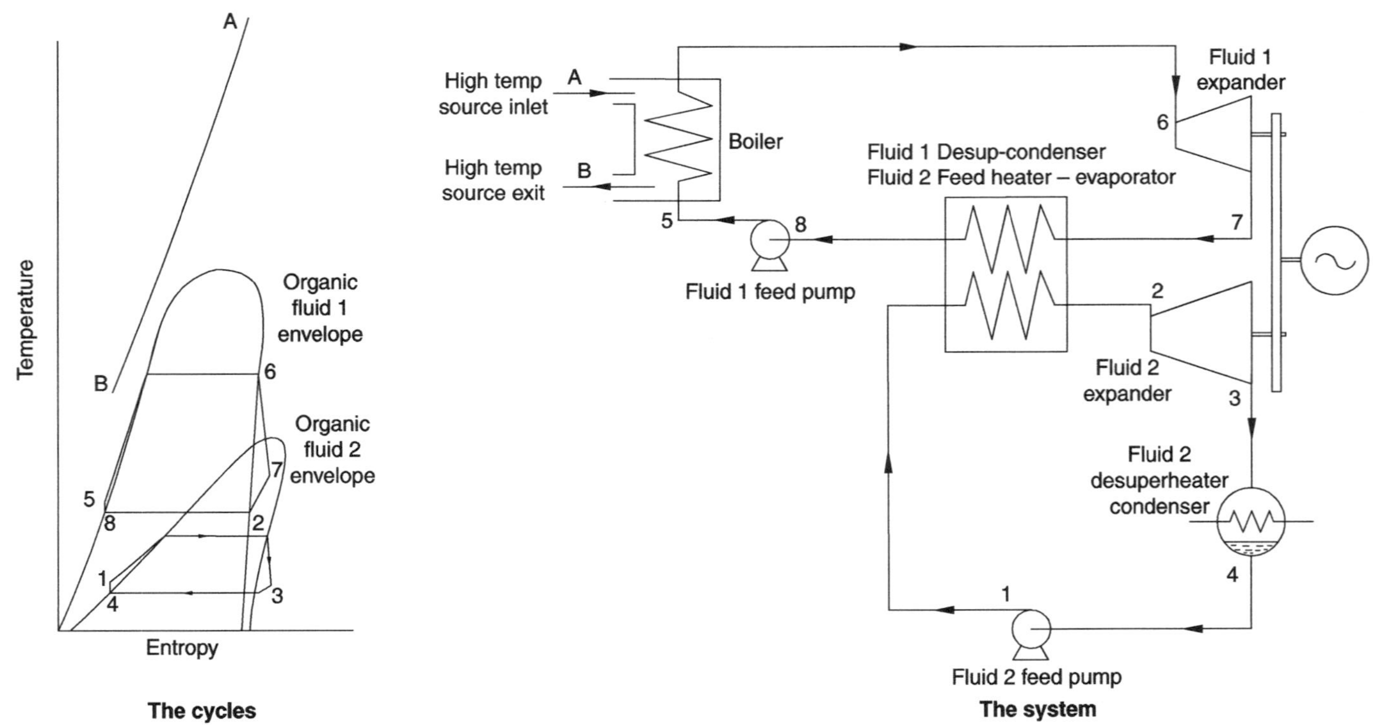

Figure 5.39 Binary cycle system with different working fluids in the topping and bottoming cycles

relatively small due to the very high heat transfer coefficients associated with condensation and boiling, while the additional cost of two separate feed pumps, compared with a single two-stage feed pump, will not be large compared with the cost of the total system.

The advantages are considerable in that a low critical temperature working fluid can be used for the bottoming cycle, thereby reducing the size and cost of the main components while coupling the two cycles in this manner enables the topping cycle to attain higher temperatures. The overall result is therefore a system that is more efficient than that based on a dual pressure cycle.

5.3.2 Wet steam cycles

The advantages of organic working fluids over steam are greatest when recovering power from heat sources in the 100–200 °C range. At higher temperatures, problems of thermal stability arise with refrigerants, especially if they contain even small amounts of lubricating oil, while plain hydrocarbons, though more thermally stable, are highly flammable. The latter factor is especially important when recovering power from IC engine exhaust gases in the 350–500 °C temperature range. At temperatures above this range, traditional steam plant is preferable, as is the practice in utilising industrial gas turbine exhaust gas heat, in combined cycle systems. Unfortunately, for maximum heat recovery in this lower temperature range, only low evaporation temperatures are possible using steam, due to its relatively large enthalpy of vaporisation, as shown in Fig. 5.40.

Organic working fluids can still be used but an intermediate thermal loop is required between the exhaust gas and the ORC boiler, as shown in Fig. 5.41, and this increases the

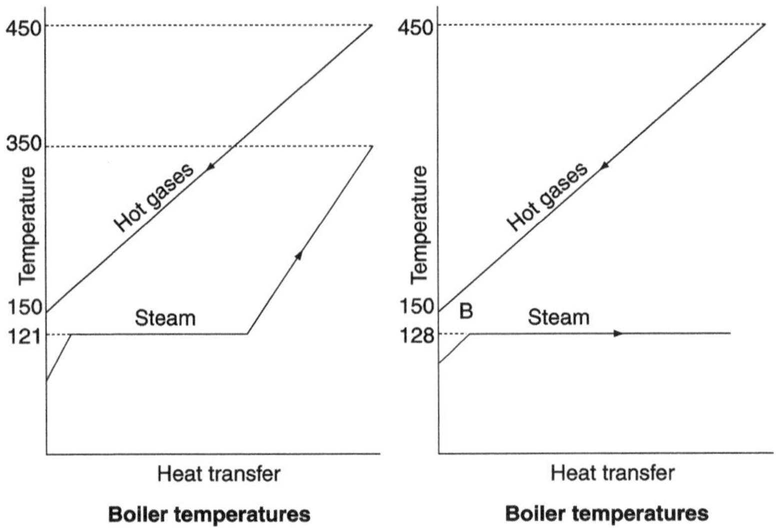

Figure 5.40 Superheated and saturated steam evaporative temperatures with maximum exhaust gas heat recovery

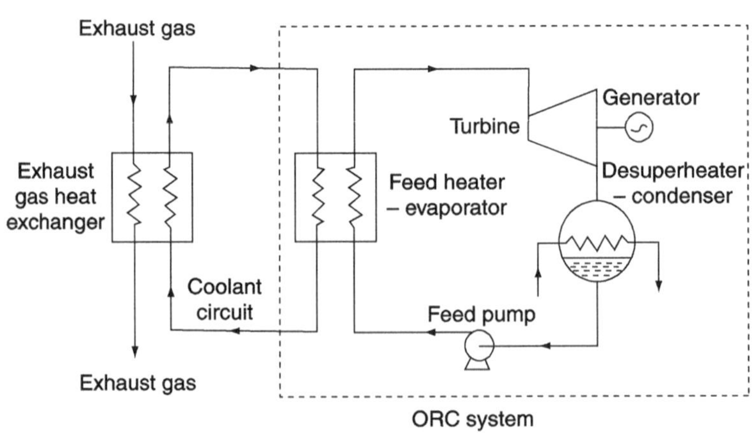

Figure 5.41 ORC system for exhaust gas heat recovery with intermediate thermal loop

system cost while decreasing the recoverable power, by limiting the maximum working fluid temperature.

Alternatively, thermal fluids with high critical temperatures can be used but these expand to sub-atmospheric conditions at normal condensing temperatures. Totally sealed systems are then required to avoid inward leakage of air and these lead to high capital costs.

An alternative approach to exhaust gas heat recovery is possible by taking full advantage of the ability of screw expanders to expand two-phase vapours. By this means it is possible to use wet steam as the working fluid with a direct exhaust gas heat exchanger and thereby attain much higher evaporative temperatures, with maximum heat recovery, at working pressures that can be maintained by normal bearings, as shown in Fig. 5.42. Typically, maximum pressures of up to 30 bar can be considered, which have corresponding saturation temperatures of up to 234 °C, which is higher than is possible with most organic fluids.

The disadvantage of condensing at-sub atmospheric pressure can be overcome by condensing at temperatures no lower than 100 °C. This will reduce the attainable efficiency of the system but not as much as it would with organic working fluids, because more heat is input at higher temperatures.

Other advantages to the use of wet steam include:

1. A much lower feed pump power is required with water as the working fluid.
2. There is no desuperheat at the end of expansion and hence no need for recuperation.
3. By condensing at 100 °C, the system will operate with air cooling throughout the year with minimum seasonal variation in power output.

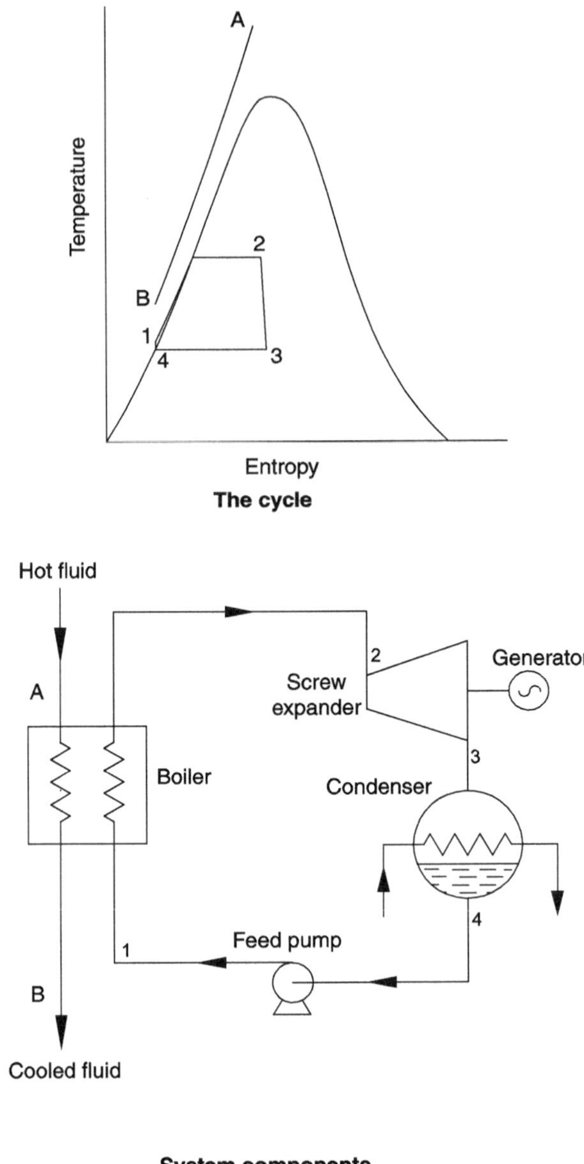

Figure 5.42 Wet steam cycle for exhaust gas heat recovery

4. In air-cooled systems, the size and cost of the condenser are drastically reduced.
5. The condensate can be used directly as the heating medium in combined heat and power (CHP) systems.
6. The size of the expander is comparable to that of a system working with an organic working fluid.

To obtain a good expander efficiency over the large pressure range required, two-stage expansion is needed and by this means overall cycle efficiencies of the order of 15% are achievable with steam of approximately 50% dryness, the latter value varying, depending on the limiting exhaust gas temperatures. Although this is less than the best possible with organic working fluids, the simplicity of the system and its low component costs give such an arrangement a lower cost per unit output than any currently known ORC system required for the same purpose.

Where abundant cooling water is available and there is no requirement for heating services from the condensate, two other system arrangements are possible.

Combined wet steam–ORC system

A combined wet steam–ORC system is shown in Fig. 5.43. Typically, by this means, overall cycle efficiencies of 20–25% are possible.

Combined screw–turbine wet steam system

Where conditions justify it, the wet steam is expanded to a higher exit pressure in the screw expander, typically of the order of 5–10 bar. The water and steam components are then separated and the dry steam is further expanded in a turbine, while the separated water is used to preheat the partially pressurised feed water, as shown in Fig. 5.44. The

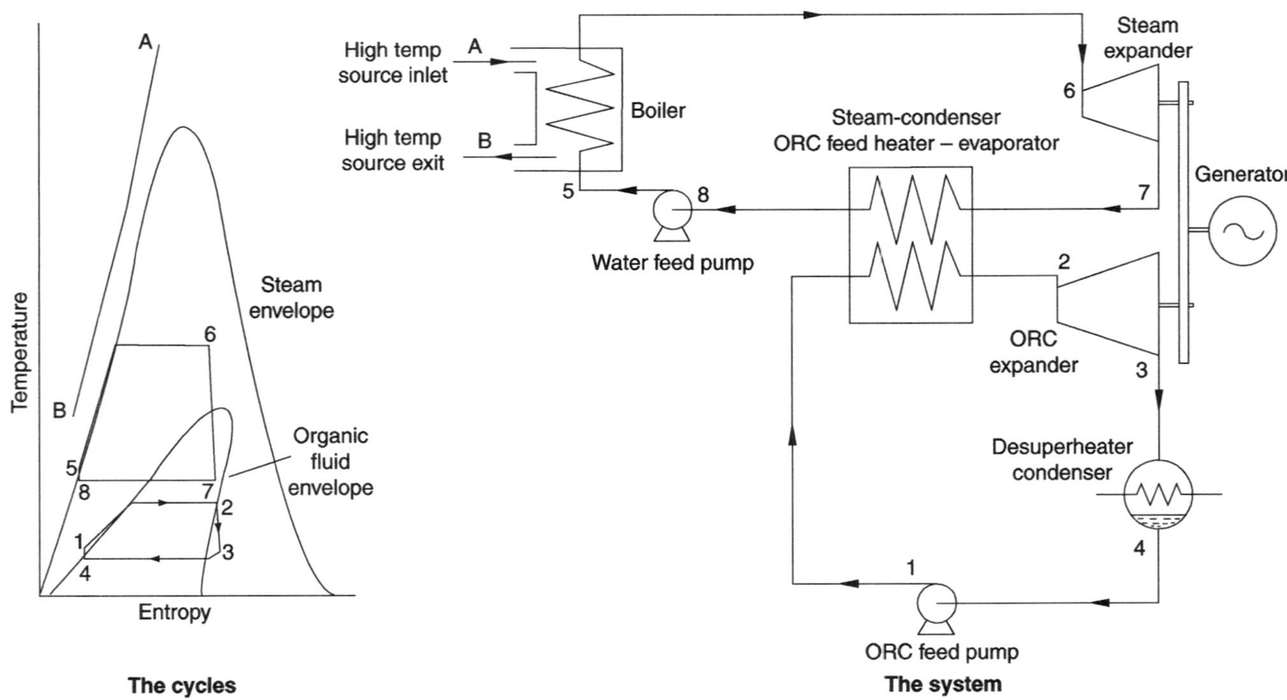

Figure 5.43 Combined wet steam–ORC system

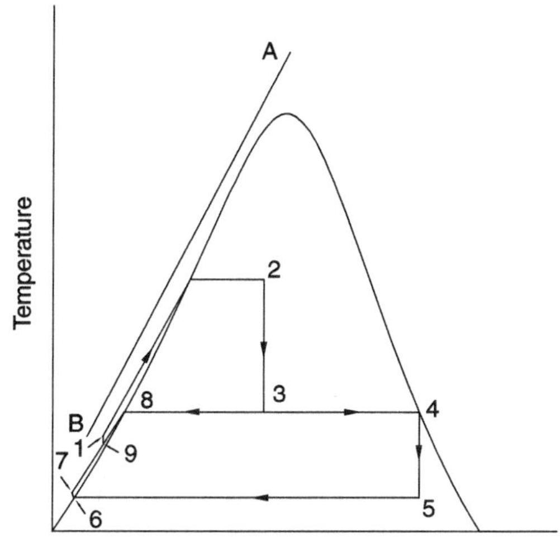

The cycle

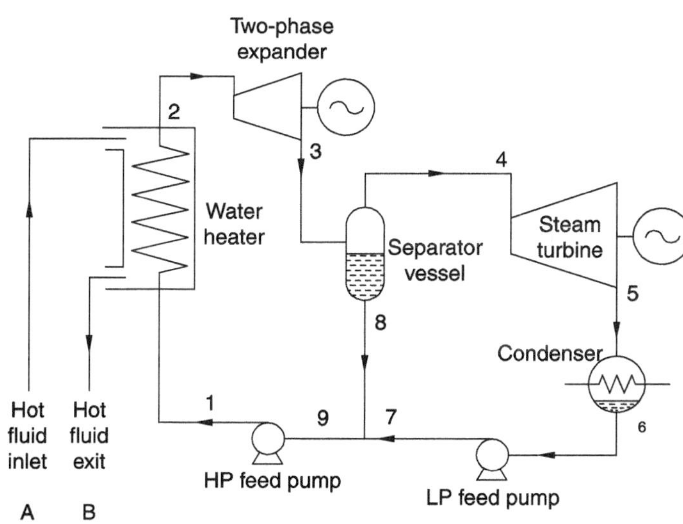

The system

Figure 5.44 Combined screw–turbine wet steam system

disadvantage of this is that a deaerator and associated components are required to maintain the vacuum at the turbine exhaust, but the advantages are that the expansion in the screw can be achieved in a single stage while an overall cycle efficiency of the order of 25% or more is possible.

Approximately the same configuration can be used to maximise power recovery from higher temperature EGS systems, as described on pp. 156 and 158, where downhole temperatures of up to 280 °C have been recorded. In such cases, the system efficiency can be further improved by creating more steam by further flashing the water leaving the separator to a lower intermediate pressure, as shown in Fig. 5.45.

In principle, the pressure reduction valve between the two separators could be replaced by a further screw expander in order to recover more power. However, in practice, due to the relatively low pressure and small pressure difference over which such an expander would operate, such an additional expander would be disproportionally large and therefore is unlikely to be cost-effective.

5.4 Cycle optimisation

In view of the relatively low efficiencies of systems for the recovery of power from low-grade heat and the fact that the temperature limits at which the systems operate are not limited by metallurgical considerations, it is vital to optimise the system design to obtain the best performance for the least cost. To this end, optimisation procedures must include not only the cycle itself but also its interaction with its auxiliary equipment. The following factors must therefore be included in any design procedure.

As described in Chapter 1, there has to be a compromise between heat recovery efficiency and cycle efficiency in order

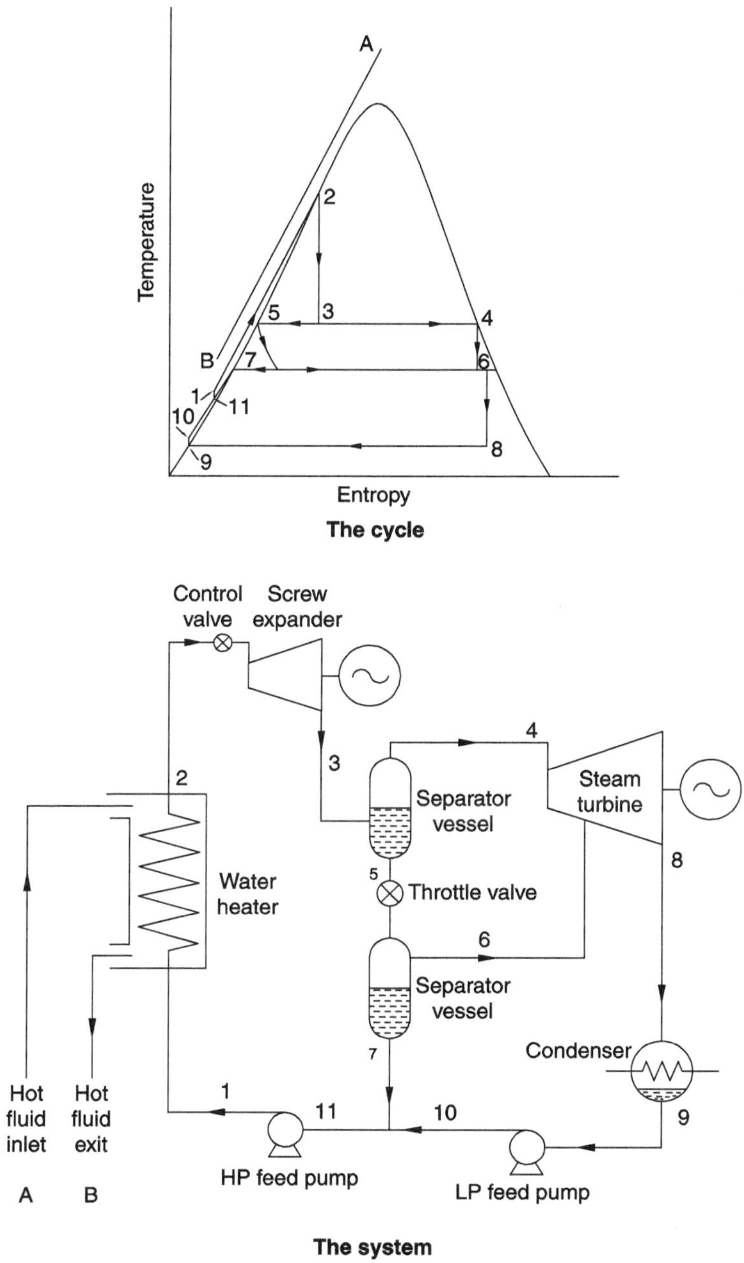

Figure 5.45 Enhanced wet steam cycle system for geothermal applications

to maximise the overall conversion efficiency or power output. It should be noted that, especially in the case of single-phase heat sources, such as hot gases or liquids, higher cycle efficiencies can be obtained by raising the maximum working fluid temperature and pressure but only at the expense of higher heat source exit temperatures from the primary heat exchanger and hence reduced heat recovery. It should be appreciated that the power output is directly proportional to the heat recovered but the gain in cycle efficiency with increased maximum working fluid temperature is not linear. Hence, in most cases, the optimum matching point is close to that associated with maximum permitted heat recovery.

Reducing the condensing temperature raises the cycle efficiency and hence the power output, but, given that the coolant temperature at the condenser inlet is fixed, this can only be achieved by reducing the temperature rise of the coolant and hence increasing its mass flow. This rises very rapidly as the temperature rise of the coolant decreases so, as the condensing temperature is reduced, a point is reached where the gain in power from the expander-generator is less than the increased power requirement to drive the condenser pump or fan. Accordingly, account of the condenser fan or coolant power must be included in the study. In initial studies, this can be accounted for by assuming a fixed value for the pressure drop of the coolant passing through the condenser, but in more advanced studies, full account of the pump or fan drive operating characteristics, together with those of its drive motor, should be included in the analysis.

The expander operating characteristics must be included in the analysis. For initial studies, it is sufficient to assume a fixed value for the expander efficiency. However, in more advanced studies, more details of the expander performance must be included. This requires an initial assumption as to what size of machine is to be used. In the case of screw expanders, the operating characteristics are dependent on the expander, length to diameter (L/D) ratio, the high-pressure port size (built in volume ratio) and the rotational speed. These must all be accounted for in the study and, in view of the fact that the screw expander has quite a limited range of built-in volume ratios over which it is efficient, it is possible that the best system design may result in a maximum fluid temperature and pressure below that initially assumed, in order to obtain the best compromise between the ideal cycle efficiency and the practical expander efficiency.

Reducing the pinch point temperature differences in the heat exchangers increases the overall performance of the system but at the expense of the heat transfer area and hence the heat exchanger cost. As a good first assumption, one may assume that in gas-heated and air-cooled systems, the overall heat transfer coefficient in the feed heater, evaporator, recuperator and condenser are all of a roughly equal value. Hence, in initial studies, a good guide to the significance of the heat exchanger size is obtained by dividing the heat transferred in the heat exchanger by the log mean temperature for the assumed conditions (Q/LMTD) and then summing up the total. This is particularly useful when comparing the relative merits of different working fluids. However, in more advanced studies, more detailed attention should be given to the heat transfer characteristics of the heat exchangers and, where possible, costing parameters for the heat exchangers should be included. As a rough guide, the general practice in initial studies is to assume a 5 °C pinch point difference in liquid-heated or cooled-heat exchangers and 10 °C for air- or gas-heated or -cooled systems.

Computer programs can be written to estimate the performance of any given system, given initial starting values to all the parameters mentioned above to produce an estimate of the overall system performance. The program can then include a multi-variable minimisation routine, such as MINUIT, to obtain the best system design. However, it should be noted that some experience is needed in selecting the initial values for the design, because the number of local minima increases with the number of variables to be optimised. Hence, initial assumptions taken too far from the final value may lead to convergence at a local optimum, which may not be for the best.

The final goal of optimisation studies should not be to maximise the power output but to minimise the total system capital cost per unit output. Detailed costing parameters are then needed for all the components. Preparing a computer program to include all these parameters is, at present, not feasible, especially when the aim is to devise a method of analysis that can be used for a wide range of applications. Ultimately, therefore, final studies will need to be carried out by trial-and-error methods in parallel with computer-aided design studies.

Appendix 1: Properties of commonly used working fluids

Refrigerant code number	Fluid	Other chemical name	Chemical formula	Molecular weight	Critical pressure (bar)	Critical temperature (°C)	Atmospheric boiling point (°C)
R123		Dichlorotrifluoroethane	$CHCl_2CF_3$	152.93	36.62	183.7	27.8
R124		Chlorotetrafluoroethane	$CHClFCF_3$	136.48	36.15	122.2	–11.2
R134a		Tetrafluoroethane	CH_2FCF_3	102.03	40.59	101.1	–26.1
R245fa		1,1,1,3,3-Pentafluoropentane	$CF_3CH_2CHF_2$	134.05	36.51	154.0	15.1
	n-Pentane		C_5H_{12}	72.15	33.70	196.6	36.1
	Neopentane	2,2-Dimethyl propane	C_5H_{12}	72.15	31.96	160.6	9.5
	Isopentane	2-Methyl butane	C_5H_{12}	72.15	33.78	187.2	27.8
R600	*n*-Butane		C_4H_{10}	58.12	37.96	152.0	–0.5
R600a	Isobutane	2-Methyl propane	C_4H_{10}	58.12	36.29	134.7	–11.7
	n-Hexane		C_6H_{14}	86.18	30.34	234.7	68.7
	Toluene	Methyl benzene	$C_6H_5(CH_3)$	92.14	41.26	318.6	110.6
R718	Water		H_2O	18.02	220.64	374.0	100.0
R744	Carbon dioxide		CO_2	44.1	73.8	31.1	–78.5

Appendix 2: Estimation of working fluid properties

Thermodynamic properties of pure fluids and their mixtures are obtained by use of appropriate property equations incorporated into the related software. These are IIR (International Institution of Refrigeration) Routines and the THERPROP databank and subroutine package developed at City University, London, as well as NIST (National Institute of Standards) property routines. Following Smith and Pitanga Marques (1993), in THERPROP, the thermodynamic properties of non-polar and weakly polar fluids were estimated with the Lee-Kesler vapour pressure equation and equation of state. Polar fluid properties were estimated with the Martin-Hou vapour equation with suitable liquid phase correlations together with the Cox-Antoine vapour equation.

Both these methods are highly accurate and revealed maximum differences of the order of ±1% in locally computed property values when compared with independent calculations carried out by other investigators using alternative procedures, such as the Starling version of the Benedict, Webb and Rubin (1940, 1942) equation (BWR equation). In the following, we describe a method developed for estimating the thermodynamic properties of mixtures, as developed by the authors, some years ago.

A2.1 The Redlich-Kwong-Soave equation of state

The Redlich-Kwong-Soave (RKS) equation of state (Soave, 1972) can be written as:

$$p = \frac{RT}{V-b} - \frac{a(T)}{V(V+b)}. \qquad [A1]$$

Substituting:

$$\frac{ZRT}{p} = V, \quad A = \frac{ap}{R^2T^2} \text{ and } B = \frac{bp}{RT}. \qquad [A2]$$

This can be rewritten in terms of the compressibility factor Z as:

$$Z^3 - Z^2 + (A - B - B^2)\,Z - AB = 0. \qquad [A3]$$

An advantage of a cubic equation of state of this type is that its roots can all be derived algebraically and hence, given p and T, V can be obtained without the need for numerical iteration procedures.

In the case of a mixture of fluids, for each component i at its critical point, the intermolecular parameter a and the co-volume b are expressed as follows:

$$a_i(T_{ci}) = \frac{\Omega_a R^2 T_{ci}^2}{p_{ci}} \text{ and } b_i(T_{ci}) = \frac{\Omega_b R T_{ci}}{p_{ci}}, \qquad [A4]$$

where the values of the constants Ω_a and Ω_b are given as:

$$\Omega_a = \frac{1}{9(2^{\frac{1}{3}}-1)} \text{ and } \Omega_b = \frac{(2^{\frac{1}{3}}-1)}{3}. \qquad [A5]$$

Soave noted that a successful correlation of phase equilibria of mixtures implied the correlation of the vapour pressures of pure substances. He defined a dimensionless function $a(T_r)$ as follows:

$$a(T_r) = a(T)/a(T_c) \quad [A6]$$

where $a(T_r) \rightarrow 1$ as $T \rightarrow T_c$.

For non-polar and slightly polar fluids, this function can be expressed in a linear form as:

$$\sqrt{\alpha} = 1 + m(1 - T_r) \quad [A7]$$

where m is a function of the acentric factor ω.

This function was originally defined by Soave (1972), but later correlated more accurately by Graboski and Daubert (1979) as:

$$m = 0.48508 + 1.55171\ \omega_i - 0.15613\ \omega_i^2 \quad [A8]$$

The acentric factor was originally defined by Pitzer and colleagues (Pitzer *et al.*, 1955a, 1955b; Pitzer and Curl, 1957). The more recent definition of Lee and Kesler (1975) is widely used and is given as:

$$\omega = \frac{\ln p_{br} - 5.92714 + \frac{6.09648}{T_{br}} + 1.28862 \ln T_{br} - 0.169347 T_{br}^6}{15.2518 - \frac{15.6875}{T_{br}} - 13.4721 \ln T_{br} + 0.43577 T_{br}^6} \quad [A9]$$

Soave's method includes classical mixing rules for the determination of vapour–liquid equilibrium parameters when mixtures are non-polar and weakly polar. These contain an adjustable binary interaction parameter (k_{ij}):

$$a = \sum_{i=1}^{n} \sum_{j=1}^{n} x_i\, x_j\, a_{ij}$$

where

$$a_{ij} = \sqrt{a_i a_j}(1 - k_{ij}) \quad [A10]$$

$$b = \sum_{i=1}^{n} x_i b_i . \qquad [A11]$$

A2.1.1 Vapour–liquid equilibrium calculation

Vapour–liquid equilibrium implies the equality of fugacities. Hence:

$$f_i^l = f_i^v \ (i = 1, \ldots\ldots n \text{ components}) \qquad [A12]$$

The fugacities are normally replaced by the fugacity coefficient, since it is more directly related to the measurable properties pressure, temperature and mole fraction. Equation [A12] then becomes

$$\phi_i^l\,(T, p, x_i)\; x_i = \phi_i^v\,(T, p, y_i)\; y_i \qquad [A13]$$

Fugacity coefficients are derived analytically and for Soave's equation of state can be expressed as follows:

$$\ln\phi_i = \frac{b_i}{b}(z-1) - \ln(z-B) + \frac{A}{B}\left[\frac{b_i}{b} - \frac{2}{a}\sum_j y_j a_{ij}\right]\ln\left(1 + \frac{B}{z}\right) \qquad [A14]$$

A2.2 Thermodynamic property estimation

The equations used for evaluating thermodynamic properties were as follows:

A2.2.1 Enthalpy

$$H = H^{o} - \Delta H + H_{Correc} \qquad [A15]$$

The first term on the right-hand side of Eq. [A15] is obtained from the ideal gas heat capacities of the pure components. These are a function of temperature and are normally expressed in a polynomial form as:

$$C_p^o = A + BT + CT^2 + DT^3 + \ldots.. \qquad [A16]$$

where

$$A = \sum_i x_i A_i \qquad B = \sum_i x_i B_i \qquad \text{etc.,} \qquad [A17]$$

so that:

$$H^o = \int_{T_o}^{T} C_p^o dT \qquad [A18]$$

Substituting for the ideal gas heat capacity, Eq. [A18] becomes:

$$H^o = A(T - T_o) + \frac{B}{2}(T^2 - T_o^2) + \frac{C}{3}(T^3 - T_o^3) + \frac{D}{4}(T^4 - T_o^4) \qquad [A19]$$

The values for A, B, C and D were obtained from the THERPROP databank.

ΔH, the enthalpy departure function, for the RKS equation is given as:

$$\Delta H = RT\left[(1 - z) + \frac{A}{B}\left(1 + \frac{D}{a}\right)\ln\left(1 + \frac{B}{z}\right)\right] \qquad [A20]$$

where

$$D = \sum_i \sum_j y_i y_j m_j (1 - k_{ij})\sqrt{a_1 \alpha_i}\sqrt{a_j Tr_j} \qquad [A21]$$

H_{Correc} sets a common base value (h_f = 100 kJ/kg at 0 °C).

A2.2.2 Entropy

$$S = S^o - \Delta S + S_{Correc} + S_{Mix}$$

where

$$S^o = A\ln\left(\frac{T}{T_o}\right) + B(T - T_o) + \frac{C}{2}(T^2 - T_o^2) + \frac{D}{3}(T^3 - T_o^3) + R\ln\left(\frac{p}{p_o}\right) \quad \text{[A22]}$$

ΔS is the entropy departure function, corrected from the given value as (Walas, 1985):

$$\Delta S = -R\ln(z - B) + \frac{AD}{Ba}\ln\left(1 + \frac{B}{z}\right) \quad \text{[A23]}$$

S_{Mix}, the entropy of mixing, is given by:

$$S_{Mix} = -R\ Y_i\ lny_i \quad \text{[A24]}$$

and S_{Correc} sets a common base value (s_f = 1.0 kJ/kg at 0 °C).

A2.2.3 Computational techniques

Equations of state can be represented graphically on pressure–volume coordinates as a family of isotherms, as shown in Fig. A1. In the case of fluid mixtures, unlike pure fluids, dew and bubble temperatures at the same vapour pressure are not normally equal. The mode of estimation of saturation conditions may be shown most simply when the temperature is given and the corresponding pressure is required. The equation is then reduced to the case of a single isotherm. A cubic equation, such as the RKS, yields three roots in the two-phase region. The largest of these corresponds to the vapour phase volume, the smallest to that of the liquid

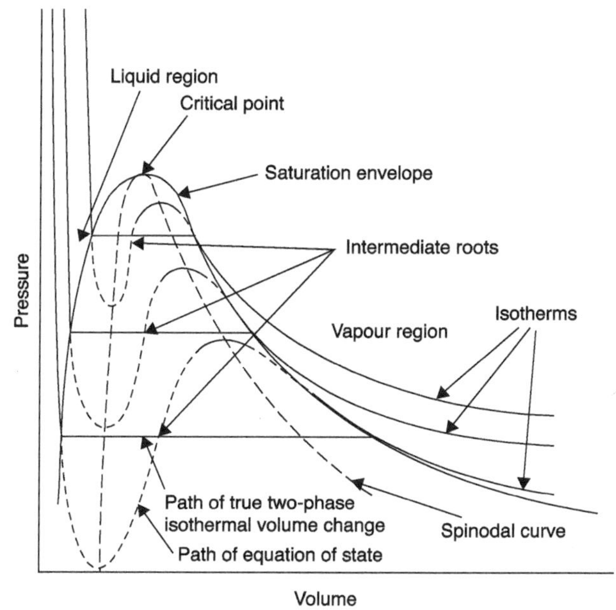

Figure A1 **Pressure–volume diagram for a fluid in the two-phase region with isotherms derived from a cubic equation of state**

phase, while the intermediate root, which is located on a positive gradient, has no physical significance. The pressure is iterated until equal fugacity values are obtained from the roots corresponding to liquid and vapour.

During the process of iteration, it is possible that values of pressure may be derived for which only one real root exists, as in the pure liquid and dry vapour regions shown in Fig. A1. In such cases, to enable the solution to continue until the required two roots are attained, the method of solution must contain procedures:

i. to determine whether the single root is a liquid or vapour value;

ii. to assign a hypothetical real root to the other phase for the fugacity comparison.

Various techniques have been used for this purpose (e.g. Poling *et al.*, 1981; Gundersen, 1982; Ziervogel and Poling, 1983). The following method was used in this investigation.

The locus of the maxima and minima of all the subcritical isotherms was obtained. This is shown in Fig. A1. For all cubic equations of state, it is known as the van der Waals spinodal curve and for the RKS equation it has the form:

$$V^4 + 2\left[b - \frac{a}{RT}\right]V^3 + \left[b^2 + \frac{3ab}{RT}\right]V^2 - \frac{ab^3}{RT} = 0 \qquad \text{[A25]}$$

The roots of this were established for the isotherm under consideration. These are fixed in bubble and dew pressure calculations but vary with each iteration in bubble and dew temperature estimation procedures. The van der Waals spinodal is a fourth-order polynomial in volume. Hence it has four roots. Imaginary and negative roots, which have no thermodynamic significance, were discarded. Since the repulsive term in any cubic equation leads to a singularity at $V = b$, it was found that only roots greater than co-volume b could be used. The single root of the RKS equation was then compared with the remaining spinodal roots. If it was larger than the biggest spinodal root, it was assumed to be a vapour root. The smallest spinodal root was taken as the liquid root. If the RKS root was smaller than the smallest spinodal root, it was assumed to be a liquid root and the largest spinodal root was taken as the vapour value.

Other problems include the selection of suitable starting values for the iterative procedure. At low pressures, the assumption of Raoult's law was sufficient for a first estimate but at higher pressures this did not lead to convergence. A variety of checks and modifications therefore had to be applied to the starting value before iteration began, in order to obtain equal fugacities. The method of iteration then used was the Newton-Raphson procedure.

Bibliography

Adam, A.W. and Monahan, J. (1973) 100 kW organic Rankine cycle total energy system, *Proceedings 8th IECEC*, pp. 126–130.

Adam, A.W., Niggemann, R.E. and Sibert, L.W. (1968) Thermal stability determination of biphenyl and the eutectic of biphenyl and phenyl ether in a Rankine cycle system, *Proceedings 3rd IECEC*, pp. 398–406.

Ainger, A. (1830) Quarterly Journal of Science, Literature and Arts, 7 (NS): 186–189.

Alaska Center for Energy and Power (2010) *Test Evaluation of Organic Rankine Cycle Engines Operating on Recovered Heat from Diesel Engine Exhaust*. Fairbanks, AK: Alaska Center for Energy and Power.

Alcock, J.F. and MacNair, E.J. (1953) High temperature vapour cycles for heat engines, *Engineering*, 175: 325.

Aljundi, I.H. (2011) Effect of dry hydrocarbons and critical point temperature on the efficiencies of organic Rankine cycle, *Renewable Energy*, 36(4): 1196–1202.

Allen, J.C.J. (undated) A proposal to combine the gas turbine and Rankine cycle plant to provide a cheap peak load generation, *CEGB Report RD/A/Rl6*.

Amelio, L. (1955) A steam engine using a mixture of vapours from non-miscible fluids as a solar engine with flat plate collectors, *Transactions Conference on Uses of Solar Energy*, 2 (Pt. 1, Section B): 238–252.

Amosov, P.E., Bobrikov, N.I., Svarc, I.I. and Vernii, A.L. (1977) *Vintovie KompresornieMashinii Spravochnik*

(*Screw Compression Expanders Handbook*). Leningrad: Mashinstroienie [in Russian].

Anderson, J.H. (1973) Vapour turbine cycle for geothermal power generation, in P. Kruger and C. Otte (eds.), *Geothermal Energy*, pp. 163–175. Stanford, CA: Stanford University Press.

Andreev, P.A. (1961) *Vintovie Kompressornie Mashinii* (*Screw Compression Machines*). Leningrad: SUDPROM [in Russian].

Andrew, C. (2006) Design and optimization of organic Rankine cycle solar–thermal power plants, Thesis, University of Wisconsin-Madison.

Angelino, G. and Colonna, P. (2000) Organic Rankine cycles for energy recovery from molten carbonate fuel cells, *35th Intersociety Energy Conversion Engineering Conference (IECEC)*, pp. 1–11.

Angelino, G. and Moroni, V. (1973) Perspectives for waste heat recovery by means of organic fluid cycles, *Journal of Engineering for Power*, 95: 75–83.

Anon (1856) *The Engineer*, 1: 45 (25 January).

Anon (1971a) Fluorocarbon turbine, *Mechanical Engineering*, 93(1): 40.

Anon (1971b) New experimental turbine, *Energy International*, 8(9): 48.

Anon (2000) MW turbine generators using steam-Freon cycle, *Electrical Review*, 177(5): 160–161.

Aoun, B. (2008) Micro combined heat and power operating on renewable energy for residential buildings, Thesis, Ecole de Mines de Paris.

Aoun, B. and Clodic, D. (2008) Theoretical and experimental study of an oil-free scroll type vapor expander, *Proceedings 19th International Compressor Engineering Conference*, Paper 1925. Available at: http://docs.lib.purdue.edu/icec/1925.

Arbon, I.M. (1994) The Design and Application of Rotary Twin-shaft Compressors in the Oil and Gas Process Industry. London: MEP.

Aronson, D. (1961) Binary cycle for power generation, *Proceedings American Power Conference*, 23: 261–271.

Astberg, A. (1982) Rotary Positive Displacement Fluid Compressor. GB Patent 2092676B.

Athans, J.N. and Dixon, J.K. (1964) Exhaust heat recovery systems for industrial gas turbine engines, *SAE Technical Paper 799C*.

Ayres, R.U. and McKenna, R.P. (1972) *Alternatives to the Internal Combustion Engine*. Baltimore, MD: Johns Hopkins University Press.

Badr, O., Probert, S. and O'Callaghan, P. (1985) Selecting a working fluid for a Rankine-cycle engine, *Applied Energy*, 21: 1–42.

Badr, O., Naik, S., O'Callagen, P. and Probert, S. (1990a) Expansion machine for a low power-output steam Rankine-cycle engine, *Applied Energy*, 39: 93–116.

Badr, O., O'Callaghan, P.W. and Probert, S.D. (1990b) Rankine-cycle systems for harnessing power from low-grade energy sources, *Applied Energy*, 36(4): 263–292.

Balje, O.E. (1962) A study on design criteria and matching of turbomachines, Part A: Similarity relations and design criteria of turbines, *Journal of Engineering for Power*, 84: 83–102.

Bammert, K. (1979) Rotating Piston Compressor with Parallel Meshing Rotors. DE Patent 2911415 [in German].

Barber, R.E. (1973) Small Rankine cycle total energy system for recreation vehicles: a comparison of three possible systems, *Proceedings 8th IECEC*, pp. 138–145.

Barber, R.E. (1974) Potentials of Rankine engines to produce power from waste heat streams, *Proceedings 9th IECEC*, pp. 508–514.

Barber, R.E., Bond, J.C. and Alford, E.H. (1971) Design and development of a turbine gearbox for use in automotive organic Rankine cycle system, *SAE Technical Paper 710564*.

Baries, D.W. and Leighton, G.S. (1960) Direct cycle diphenyl reactor, *Transactions American Nuclear Society*, 3: 553–554.

Baries, D.W. and Leighton, G.S. (1962) Diphenyl shows advantages for (i) direct cycle nuclear plants, (ii) dual purpose plants, *Power Engineering*, 66: 43–45.

Barkef, G. (1965) Waste heat power recovery for a gas turbine engine, MSc thesis, Imperial College, University of London.

Barret, R.W., Garcia, R. and Waters, W.M. (1970) Organic Rankine power unit testing, *Proceedings 5th IECEC*, pp. 11.62–11.66.

Bein, T.W. and Hamilton, J.F. (1982) Computer modeling of an oil flooded single screw air compressor, *Proceedings International Compressor Engineering Conference*, Paper 383. Available at: http://docs.lib.purdue.edu/icec/383.

Beldecos, N.A. (1955) Benzene turbine cycles, *USAEC Report WIAP-M-31*, June.

Beldecos, N.A. and Lokay, J.D. (1953) Thermodynamic investigations of the benzene turbine cycle, *USAEC Report WIAP-M-24*, July.

Benedict, M., Webb, G.B. and Rubin, L.C. (1940) An empirical equation for thermodynamic properties of light hydrocarbons and their mixtures. I. Methane, ethane, propane and butane, *Journal of Chemical Physics*, 8: 334–345.

Benedict, M., Webb, G.B. and Rubin, L.C. (1942) An empirical equation for thermodynamic properties of light hydrocarbons and their mixtures. II. Mixtures of methane, ethane, propane and *n*-butane, *Journal of Chemical Physics*, 10: 747–758.

Benedict, M., Webb, G.B. and Rubin, L.C. (1951a) An empirical equation for thermodynamic properties of light hydrocarbons and their mixtures: constants for twelve hydrocarbons, *Chemical Engineering Progress*, 47(8): 419–422.

Benedict, M., Webb, G.B. and Rubin, L.C. (1951b) An empirical equation for thermodynamic properties of light hydrocarbons and their mixtures: fugacities and liquid–vapour equilibria, *Chemical Engineering Progress*, 47(9): 449–454.

Benstein, E.H. and Wood, H.J. (1964) Applications and performance levels of radial inflow turbines, *Transactions SAE*, 72: 492–501.

Bertrand, T.F., Papadakis, G., Lambrinos, G. and Frangoudakis, A. (2009) Fluid selection for a low-temperature solar organic Rankine cycle, *Applied Thermal Engineering*, 29(11/12): 2468–2476.

Bindon, J.P. (1969) Fluid flow in high mach no. axial flow turbines, PhD thesis, Imperial College, University of London.

Bindon, J.P. and Charmichael, A.D. (1968) A preliminary assessment of the combined gas turbine and vapour turbine engines, *Journal of the Institute of Fuel*, 41(324): 28–36.

Bjerklie, J.W. (1967a) Criteria for minimum weight Carnot limited space power systems, *Proceedings 2nd IECEC*, pp. 123–133.

Bjerklie, J.W. (1967b) Working fluid as a design variable for a family of small Rankine power systems, *ASME Paper 67-97-6*.

Bjerklie, J. and Luchter, S. (1969) Rankine cycle working fluid selection and specification rationale, *SAE Technical Paper 690063*.

Bjerklie, J.W. and Sternlicht, B. (1969) Critical comparison of low emission Otto and Rankine engines for automotive use, *SAE Technical Paper 690044*.

Bjerklie, J.W. and Sternlicht, B. (1970) Where closed Rankine cycles fit – with examples of automotive APS and portable GPS, *SAE Technical Paper 700120*.

Blake, E.L., Hamman, W.E., Edwards, J.W., Reichard, L.E. and Art, M.R. (1961) Thermal stability as a function of chemical structure, *Journal of Chemical and Engineering Data*, 6(1): 87–98.

Boretz, J.E. (1967) Isotopic dynamic space power systems, *Proceedings 2nd IECEC*, pp. 615–651.

Boretz, J.E. (1972) Low peak temperatures and hydrodynamic bearings – key to long life organic Rankine cycle systems, *Proceedings 7th IECEC*, pp. 296–302.

Boretz, J.E. (1973) Isotope organic Rankine cycle electric power systems for the 150–1500 watt range, *Proceedings 8th IECEC*, pp. 601–608.

Boretz, J.E. (1974) Technology considerations for organic Rankine cycle electric power systems, *Proceedings 9th IECEC*, pp. 61–67.

Boretz, J.E., Lurie, H., Wright, H., Nagyszalanczy, L. and Silver, A. (1974) Turbine organic Rankine cycle engine system for artificial heart application, *Proceedings 9th IECEC*, pp. 813–823.

Bourne, J. (1861) *A Treatise on the Steam Engine* (5th edn.), pp. 393–394. London: Longman, Green, Longman, Roberts.

Bowman, J.L. (1983) US Patent 4,412,796.

Brasz, J.J., Shistla, V., Stosic, N. and Smith, I.K. (2000) Development of a twin screw expressor as a throttle valve replacement for water-cooled chillers, *Proceedings XVth International Compressor Engineering Conference*, July.

Brasz, J., Biederman, B. and Holdmann, G. (2005) Power production from a moderate-temperature geothermal resource, *Geothermal Resources Council Transactions*, 29: 25–28 September.

Brogan, J.J. and Thur, G.M. (1972) Advanced automotive power systems development program, *Proceedings 7th IECEC*, pp. 806–824.

Broglia, T., Iobbi, M. and Stosic, N. (2006) Improving the capacity and performance of air-conditioning screw compressors, *IMECE 2006, ASME Congress*, pp. 37–42. Chicago, IL, November.

Bronicki, L. (1966) Rankine cycle power unit operating with isotopic heat sources, *UKAEA/ENEA Symposium on Industrial Applications for Isotopic Power Generators*, Harwell, September.

Bronicki, L.Y. (1972) The Ormat Rankine power unit, *Proceedings 7th IECEC*, pp. 327–334.

Bronicki, L. and Tabor, H.Z. (1965) Tiny turbine plus right working fluid equals high efficiency, *SAE Journal*, 73(6): 52–55.

Brown, F.R.S. (1966) The prospects of alternative methods of generation of electrical power: a comprehensive review, *Transactions World Power Conference*, 3: 1441–1453.

Bruno, J., Lopez-Villada, J., Letelier, E., Romera, S. and Corona, A. (2008) Modelling and optimization of solar organic Rankine cycle engines for reverse osmosis desalination, *Applied Thermal Engineering*, 28: 2212–2226.

Bryson, M. (2007) The conversion of low-grade heat into electricity using the thermo siphon Rankine engine and trilateral flash cycle, Thesis, Royal Melbourne Institute of Technology.

Bundela, P.S. and Chawla, V. (2010) Sustainable development through waste heat recovery, *American Journal of Environmental Sciences*, 6(1): 83–89.

Burnside, B.M. (1973) Thermodynamic properties of five halogenated hydrocarbon vapour power cycle working fluids, *Journal of Mechanical Engineering Science*, 15(2): 132–143.

Buxton, O.E. (1969) Using the Rankine cycle for fluid evaluation, *Proceedings 4th IECEC*, pp. 614–623.

Canada, S., Cohen, G., Cable, R., Brosseau, D. and Price, H. (2005) Parabolic trough organic Rankine cycle solar power plant, *Conference Paper NREL/CP-550-37077*.

Casci, C. and Angelino, G. (1969) The dependence of power cycles performance on their location relative to the Andrews curve, *ASME Paper 69-GT-65*.

Chammas, R. (2005) Rankine cycle for hybrid vehicles: simulation and design of a first prototype, Thesis, Ecole des Mines de Paris.

Chammas, R. and Clodic, D. (2005) Combined cycle for hybrid vehicles, *SAE Technical Paper 2005-01-1171*.

Chandramohan, S. (2008) First and second law analysis of organic Rankine cycle, Thesis, Mississippi State University.

Charlesworth, A.D., Skelton, P.G. and Treays, P. (1968) An investigation of the feasibility of a low cost 10 kW vapour cycle generator, *RMCS Shrivenham Project Study Report*, November.

Chen, C.H. (1995) Screw Compressor with Rotors Having Hyper Profile. US Patent 5454701.

Chen, Y., Lundqvist, P., Johansson, A. and Platell, P. (2006) A comparative study of the carbon dioxide transcritical power cycle compared with an organic Rankine cycle with R123 as working fluid in waste heat recovery, *Applied Thermal Engineering*, 26: 2142–2147.

Chuen-Sen, L. (2008) *Capture of Heat Energy from Diesel Engine Exhaust*, Final Report. Washington, DC: US Department of Energy,

Copenhaver, C.M. (1971) Small boiling organic reactor power plant, *Proceedings 6th IECEC*, pp. 768–774.

Crook, A.W. (ed.) (1994) *Profiting from Low-Grade Heat*. The Watt Committee on Energy Report No. 26, London, December.

Curtis, E.M. (1968a) Condensing heat transfer of Freon 21 on externally finned horizontal tubes, *CEGB Research Report RD/M/R8*, February.

Curtis, E.M. (1968b) Boiling heat transfer of Freon 21 in a vertical electrically heated test section, *CEGB Research Report RD/M/R96*, August.

Dai, Y., Wang, J. and Gao, L. (2009) Parametric optimization and comparative study of organic Rankine cycle (ORC) for low grade waste heat recovery, *Energy Conversion and Management*, 50(3): 576–582.

Danilin, V.S., Zakharow, I.I., Loginov, A.A. and Cherngaev, V.A. (1965) Some characteristics of diphenyl heating turbines and their limiting power, *Soviet Atomic Energy*, 19(3): 1172–1176. [Translation from *Atomnaya Energiya*, 19(3): 257–261.]

Davidson, T.A. (1977) Design and analysis of a 1 kW Rankine power cycle, employing a multi-vane expander, for use with a low temperature solar collector, Thesis, Massachusetts Institute of Technology.

De Vlaminck, M. (1988) Turbines axiales, turbines radials et machines de detente a vis: domains d'utilisation dans des installations ORC, *Revue generate de thermique*, 27: 18–27.

Degner, V.R. and Velie, W. (1969) Demonstration of a self contained organic Rankine silent engine, *SAE Technical Paper 690062*.

Degner, V.R., Velie, W., Bjerklie, J. and Luchter, S. (1970) Correct working fluid makes organic Rankine cycle work, *SAE Journal*, 78(6): 58–64.

Delgado-Torres, A.M. and Garcia-Rodriguez, L. (2007) Comparison of solar technologies for driving a desalination system by means of an organic Rankine cycle, *Desalination*, 216(1/3): 276–291.

Delgado-Torres, A.M. and Garcia-Rodriguez, L. (2010) Analysis and optimization of the low-temperature solar organic Rankine cycle (ORC), *Energy Conversion and Management*, 51(12): 2846–2856.

Desai, N.B. and Bandyopadhyay, S. (2009) Process integration of organic Rankine cycle, *Energy*, 34(10): 1674–1686.

Diamant, R.M. (1973) *Total Energy*. Oxford: Pergamon Press.

Diamond, S.D. and Taylor, A.M. (1969) An organic Rankine cycle for manned space applications, *Proceedings 4th IECEC*, pp. 447–455.

Doerner, W.A. and Bechtold, M.F. (1972) Rankine cycle engines reconsidered, *Innovation*, 4(1): 1–4.

Dorrance, W.H. (undated) Choosing working fluids for heat actuated dual loop systems, *American Gas Association Catalogue No. M 32548*.

Dow Chemical Co. (1963) *Dowtherm A and E Handbook*. Midland, MI: Dow Chemical Co.

Dow, H.H. (1926) Diphenyl oxide bi-fluid power plants, *Mechanical Engineering*, 48(8): 815–818.

Doyle, E. and Patel, P. (1976) Compounding the truck diesel engine with an organic Rankine cycle system, *SAE Technical Paper 760343*.

Doyle, E., Le Feuvre, T. and Raymond, R.J. (1970) Some developments in small reciprocating Rankine cycle engines using organic working fluids, *SAE Technical Paper 700162*.

Drescher, U. and Bruggemann, D. (2007) Fluid selection for the Organic Rankine Cycle (ORC) in biomass power and heat plants, *Applied Thermal Engineering*, 27(1): 223–228.

Eaves, P.S.K. and Hadrill, H.F.J. (1968) Factors affecting the application of binary cycle plant to the C.E.G.B. system, *Transactions 7th World Power Conference, Paper Cl-241*.

Eckard, S. (1975) Multi-vane expander as prime mover in low-temperature solar or waste heat applications, *Proceedings 1975 IECEC*, pp. 249–254.

Edstroem, S.E. (1974) US Patent 3,787,154.

Edstroem, S.E. (1992) A modern way to good screw compressors, *Proceedings International Compressor Engineering Conference*, Paper 832. Available at: http://docs.lib.purdue.edu/icec/832.

El Chammas, R. and Clodic, D. (2005) Combined cycle for hybrid vehicles, *SAE Technical Paper 01-1171*.

El Hassan, K. and Shouman, A.R. (1973) Steam-refrigerant binary power cycles, *ASME Paper 73-IPWR-14*.

Ellis, S.J. (1938/39) The total heat entropy diagram for diphenyl, *Journal of the Institute of Civil Engineers*, 10(2): 227–240.

El-Ramly, N. and Budenholzer, R.A. (1963) Binary cycle for nuclear power generation using steam and refrigerant gases, *Proceedings American Power Conference*, 25: 496–504.

Endo, T., Kawajiri, S., Kojima, Y., Takahashi, K., Baba, T. *et al.*, (2007) Study on maximizing exergy in automotive engines, *SAE Technical Paper 01-0257*.

Engin, T. and Ari, V. (2005) Energy auditing and recovery for dry type cement rotary kiln systems – a case study, *Energy Conversion and Management*, 46(4): 551–562.

Ennis, W.D. (1910) Applied Thermodynamics for Engineers. London: Constable.

Ennis, W.D. (1912) *Vapours for Heat Engines*. London: Constable.

Facao, J., Palmero-Marrero, A. and Oliveira, A.C. (2008) Analysis of a solar assisted micro-cogeneration ORC system, *International Journal of Low-Carbon Technologies*, 3(4): 254.

Fenton, D., Abernathy, G., Krivokapich, G. and Otts, J. (1984) Operation and evaluation of the Willard solar thermal power irrigation system, *Solar Energy*, 32: 735–751.

Findlay, W.S. (1934) Some suggestions for diphenyl heat engine cycles, *The Power Engineer*, 29(336): 89–91.

Fink, C.R. and Krueger E.C.A. (1970) 2–10 kWe radioisotope fuelled organic Rankine power system, *AIAA Paper 70-520*.

Fleming, J.S. and Tang, Y. (1994) The analysis of leakage in a twin screw compressor and its application to performance improvement, *Journal of Process Mechanical Engineering*, 209: 125.

Fleming, J.S., Tang, Y. and Cook, G. (1998) The twin helical screw compressor, Part 1: Development, applications and competitive position; Part 2: A mathematical model of the working process, *Journal of Mechanical Engineering Science*, 212: 369.

Fraas, A.P. (1973) A potassium–steam binary vapour cycle for better fuel economy and reduced thermal pollution, *Journal of Engineering for Power*, 95: 53–63.

Freymann, R., Strobl, W. and Obieglo, A. (2008) The turbosteamer: a system introducing the principle of cogeneration in automotive applications, *Motortechnische Zeitschrift*, 69: 20–27.

Fujiwara, M. and Osada, Y. (1995) Performance analysis of oil injected screw compressors and their application, *International Journal of Refrigeration*, 18(4): 220–227.

Fujiwara, M., Mori, H. and Suwama, T. (1974) Prediction of the oil free screw compressor performance using digital computer, *Proceedings International Compressor Engineering Conference, Paper 119*. Available at: http://docs.lib.purdue.edu/icec/119.

Fujiwara, M., Kasuya, K., Matsunaga, T. and Watanabe, M. (1984) Computer modeling for performance analysis of rotary screw compressor, *Proceedings International*

Compressor Engineering Conference, Paper 503. Available at: http://docs.lib.purdue.edu/icec/503.

Fukazawa, Y. and Ozawa, U. (1980) Small screw compressors for automobile air conditioning systems, *Proceedings International Compressor Engineering Conference, Paper 351*. Available at: http://docs.lib.purdue.edu/icec/351.

Gaffert, G.A. (1934) High pressure steam and binary cycles as a means of improving power station efficiency, *Transactions ASME*, 56(10): 755–765.

Gaffert, G.A. (1952) *Steam Power Stations*, 4th edn. New York: McGraw-Hill.

Gallant, R.W. (1968) *Physical Properties of Hydrocarbons*, Vol. 1. Houston, TX: Gulf Publishing.

Gallant, R.W. (1970) *Physical Properties of Hydrocarbons*, Vol. 2. Houston, TX: Gulf Publishing.

Gantt, H.L. and Haury, D.H. (1884) The efficiency of fluid in vapour machines, *Van Nostrands Engineering Magazine*, 31: 413–433.

Garcia, R. (1969) Jet condenser development for an organic Rankine cycle power conversion system, *Proceedings 4th IECEC*, pp. 11–20.

Garduza, O., Sanchez, F., Martinez, D. and Roman, R. (2002) Vapour pressures of pure compounds using Peng-Robertson equation of state with three different attractive terms, *Fluid Phase Equilibria*, 198: 195–228.

Gokhshtein, D.P. (1960) Raising the efficiency and output of gas turbines, *Teploenergetika*, 7: 33–38.

Gokhshtein, D.P., Smirnov, G.F. and Kirov, V.S. (1966) Some features of binary cycle systems with non-aqueous vapours, *Thermal Engineering*, 13: 25–30.

Golovintsov, A.G. *et al.*, (1964) *Rotatsionii kompresorii* (*Rotary Compressors*). Moscow: Mashinostroenie (in Russian).

Graboski, M.S. and Daubert, T.E. (1979) Modified Soave equation of state for phase equilibrium calculations, *Industrial and Engineering Chemistry Process Design and Develeopment*, 18(2): 300–306.

Gu, W., Weng, Y., Wang, Y. and Zheng, B. (2009) Theoretical and experimental investigation of an organic Rankine cycle for a waste heat recovery system, *Proceedings IMechE, Part A: Journal of Power and Energy*, 223(5): 523–533.

Gundersen, T. (1982) Numerical aspects of the implementation of cubic equations of state in flash calculation routines, *Computational Chemical Engineering*, 6(3): 245–255.

Gurgenci, H. (1986) Performance of power plants with organic Rankine cycles under part-load and off-design conditions, *Solar Energy*, 36: 45–52.

Guthrie, J. (1970) *Bizarre Ships of the Nineteenth Century*, pp. 193–196. London: Hutchinson.

Hackett, R.N. (1950) Mercury cycle power generation: a progress report, *Transactions 4th World Power Conference*, 3: 1585–1599.

Hanjalic, K. and Stosic, N. (1997) Development and optimization of screw compressors with a simulation model, Part II: Thermodynamic performance simulation and design optimization, *ASME Journal of Fluids Engineering*, 119: 664–670.

Hansen, A. (1961) Thermal cycles for geothermal sites and turbine installation at the Geyser Power Plant, California, *Proceedings UN Conference on New Sources of Energy*, 3: 365–379.

Harada, K.J. (2010) Development of a small scale scroll expander, MSc thesis, Oregon State University.

Harinck, J., Turunen-Saaresti, T., Colonna, P., Rebay, S. and van Buijtenen, J. (2010) Computational study of a high-expansion ratio radial organic Rankine cycle turbine stator, *ASME Journal of Engineering for Gas Turbines and Power*, 132(5): 054501.

Hartley, H. (1960) The Wilkins Lecture: Sir Humphrey Davy, *Proceedings Royal Society of London A*, 225: 153–180.

Herbert, M.V. (1969) Some notes on vapour cycles using fluids of high molecular weight to provide aircraft auxiliary power, *N.G.T.E. Turbomachinery Note 977*, January.

Hettiarachchi, H.D., Golubovic, M., Worek, W.M. and Ikegami, Y. (2007) Optimum design criteria for an organic Rankine cycle using low-temperature geothermal heat sources, *Energy*, 32(9): 1698–1706.

Hodgson, J.N. and Kreeger, A.H. (1973) Design and demonstration of low emission Rankine cycle automotive engine using organic working fluid and turbine expander, *Proceedings 8th IECEC*, addendum.

Holmes, C.S. and Stephen, A.C. (1999) Flexible profile grinding of screw compressor rotors, *Proceedings IMechE International Conference on Compressors and Their Systems*, p. 126.

Horn, G. and Norris, T.D. (1962) The steam refrigerant binary cycle for very large turbomachinery unit sizes, *CEGB Research Report RD/M/N39*.

Horn, G. and Norris, T.D. (1966) The selection of working fluids other than steam for future power generation cycles, *The Chemical Engineer*, 203: 298–305.

Horn, G., Norris, T.D. and Whybrow, J.F.T. (1968/69) Turbine flow problems of binary cycles employing high density fluids, *Proceedings IMechE*, 183 (Pt. 1): 165–178.

Hough, D. and Morris, S.J. (1984) Screw Rotor Compressors. GB Patent 2159883A.

Howard J.M. (1968) Nuclear organic Rankine/thermoelectric systems, *Proceedings 3rd IECEC*, pp. 533–538.

Hryniszak, W. (1968) Aspects of broadening the choice of heat engines in future, *Transactions North East Coast Institute of Engineers and Shipbuilders*, 84(3): 71–98.

Hryniszak, W. and Stoddart, D.E. (1972) Extending the range of working fluids for gas turbines, in S.W. Sawyer (ed.), *Sawyer's Gas Turbine Engineering Handbook, Vol. 1: Theory and Design*, 2nd edn. Norwalk, CT: Turbomachinery International Publications.

Hryniszak, W., Hutchinson, M. and Renton, A. (1966) Special fluid power plants, *ASME Paper 66-GT-84*.

Hudson, R. (1988) Technical and economic overview of geothermal atmospheric exhaust and condensing turbines, binary cycle and biphase plant, *Geothermics*, 17: 51–74.

Huffman, G.D. and Buxton, O.E. (1968) Thermodynamic optimization of Rankine cycle space power system, *Journal of Engineering for Power*, 90: 89–96.

Hung, T. (2001) Waste heat recovery of organic Rankine cycle using dry fluids, *Energy Conversion and Management*, 42: 539–553.

Hung, T., Wang, S., Kuo, C., Pei, B. and Tsai, K. (2010) A study of organic working fluids on system efficiency of an ORC using low grade energy sources, *Energy*, 35: 1403–1411.

Ichikawa, S. (1970) Use of fluorocarbon turbine in chemical plants, *Chemical Economy and Engineering Review*, October, pp. 19–26.

Imperial Smelting Corporation Ltd. (1965) *Product Data for Flutec PP Series and Hexafluorobenzene*. Organic Coolants and Moderators, Bibliographical Series No. 17. Vienna: International Atomic Energy Agency.

Invernizzi, C., Iora, P. and Silva, P. (2007) Bottoming micro-Rankine cycle for micro-gas turbines, *Applied Thermal Engineering*, 27: 100–110.

James, F. and Roos, M. (1975) MINUIT: a system for function minimization and analysis of the parameter errors and correlations, *Computer Physics Communications*, 10: 343.

Jameson, J.W. (1858/59) On the performance of the screw steamship 'Sahel' fitted with Du Trembleys combined vapour engine and of the sister ship 'Oasis' fitted with steam engines worked expansively and provided with partial surface condensation, *Proceedings Institute of Civil Engineers*, 18: 233–294.

Jamieson, D.T., Irving, J.B. and Tudhope, J.S. (1974) *Liquid Thermal Conductivity – A Data Survey to 1973*. Palo Alto, CA: Pendragon House.

Kane, E.H. (2002) Integration et optimisation thermo-economique et environomique de centrales thermiques solaires hybrids, PhD thesis, Laboratoire d'Energetique Industrielle, Ecole polytechnique Federale de Lausanne.

Kane, M., Larrain, D., Favrat, D. and Allani, Y. (2003) Small hybrid solar power system, *Energy*, 28(14): 1427–1443.

Kaplan, E. and Lodwig, E. (1969) Design and performance of a mobile waste heat organic Rankine cycle system providing electric power and environmental control, *Proceedings 4th IECEC*, pp. 484–493.

Kasuya, K. *et al.* (1983) US Patent 4,406,602.

Kaushik, S., Dubey, A. and Singh, M. (1994) Steam Rankine cycle cooling system: analysis and possible refinements, *Energy Conversion Management*, 35: 871–886.

Kestin, J., DiPippo, R., Khalifa, H.E. and Ryley, D.J. (eds.) (1980) *Sourcebook on the Production of Electricity from Geothermal Energy*. Washington, DC: US Department of Energy.

Kim, H., Ahn, J., Park, I. and Rha, R. (2007) Scroll expander for power generation from a low-grade steam source,

Proceedings IMechE, Part A: Journal of Power and Energy, 221: 705–711.

King, C.D. and Moodie, W.J. (1963) Biphenyl liquid heat transfer and vapour deposition studies at 700° and 800 °F, *USAEC Report IDO 11006*.

Kippenham, C.J. and Georgian, J.C. (1963) Binary fluid waste heat recovery cycle for engine and turbine prime movers, *ASME Paper 63-0GP-19*.

Koenig, J.B. (1973) Worldwide status of geothermal resources development, in P. Kruger and C. Otte (eds.), *Geothermal Energy*, pp. 15–67. Stanford, CA: Stanford University Press.

Konka, K.-H. (1988) *Schraubenkompressoren* (*Screw Compressors*). Dusseldorf: VDI-Verlag (in German).

Kovacevic, A., Stosic, N. and Smith, I.K. (2007) Screw Compressors – Three-dimensional Computational Fluid Dynamics and Solid Fluid Interaction. New York: Springer-Verlag.

Kovacevic, A., Stosic, N., Smith, I.K., Mujic, E. and Guerrato, D. (2008) Advances in numerical and experimental investigation of screw compressors, *Proceedings International Conference on Compression and Refrigeration*, pp. 288–298, Xian, China.

Lakew, A.A. and Bolland, O. (2010) Working fluids for low-temperature heat source, *Applied Thermal Engineering*, 30(10): 1262–1268.

Larjola, J. (1995) Electricity from industrial waste heat using high-speed organic Rankine cycle, *International Journal of Production Economics*, 41: 227–235.

Lee, B.I. and Kesler, M.G. (1975) A generalised thermodynamic correlation based on Pitzer's three parameter corresponding states principle, *AIChE Journal*, 21(3): 510–527.

Lee, H.T. (1988) Screw-Rotor Compressor with an Ellipse as a Part of its Male Rotor. US Patent 4,890,992.

Lee, M.J., Tien, D.L. and Shao, C.T. (1993) Thermophysical capability of ozone-safe working fluids for an organic Rankine cycle system, *Heat Recovery Systems and CHP*, 13(7): 409–418.

Leighton, G.S. (1968) The Organic Rankine Cycle, *Proceedings 3rd IECEC*, pp. 389–397.

Lemort, V. (2009) Contribution to the Characterization of Scroll Machines in Compressor and Expander Modes. Liege: University of Liege.

Lemort, V., Teodorese, I. and Lebrun, J. (2006) Experimental study of the integration of a scroll expander into a heat recovery Rankine cycle, *International Compressor Engineering Conference*.

Li, J., Pei, G. and Ji, J. (2010) Optimization of low temperature solar thermal electric generation with Organic Rankine Cycle in different areas, *Applied Energy*, 87(11): 3355–3365.

Li, J., Pei, G., Li, Y. and Ji, J. (2011) Evaluation of external heat loss from a small-scale expander used in Organic Rankine Cycle, *Applied Thermal Engineering*, 31: 2694–2701.

Lin, C. (2008) Feasibility of using power steering pumps in small-scale solar thermal electric power systems, SB thesis, Massachusetts Institute of Technology.

Linhardt, H.D. and Carver, G.P. (1967) Development progress of organic Rankine cycle power systems, *Proceedings 2nd IECEC*, pp. 103–115.

Lippke, F. (1995) *Simulation of the Part Load Behavior of a 30MWe SEGS Plant*. Albuquerque, NM: Sandia National Laboratories.

Litvin, F.L. (1994) *Gear Geometry and Applied Theory*. Englewood Cliffs, NJ: Prentice-Hall.

Liu, B.T., Chien, K.H. and Wang, C.C. (2004) Effect of working fluids on organic Rankine cycle for waste heat recovery, *Energy*, 29(8): 1207–1217.

Lodwrig, E. (1970) Performance of a 35 h.p. organic Rankine cycle exhaust gas powered system, *SAE Paper 700160.*

Luchter, S. (1967) A quantitative method of screening working fluids for Rankine cycle power plants, *ASME Paper 67-GT-12.*

Ludvigsen, K. (1970) Back to steam power, *Motor*, 138(3573): 20–24.

Lukawski, M. (2009) Design and optimization of standardized Organic Rankine Cycle power plant for European conditions, Thesis, University of Iceland.

Lyon, W.C. and Starr, M.D. (1967) Terrestrial radioisotope fuelled kilowatt power systems, *Proceedings 2nd IECEC*, pp. 821–828.

Lysholm, A. (1942) A new rotary compressor, *Proceedings IMechE*, 150: 11.

Lysholm, A. (1967) Screw Rotor Compressor. US Patent 3,314,598.

Mackay, D.B. and Leventhal, E.L. (1960) Solar turbo power plant design, *Journal of Engineering for Power*, 82: 315–324.

Madhawa, H., Golubovic, M., Worek, W. and Ikegami, Y. (2007) Optimum design criteria for an Organic Rankine cycle using low-temperature geothermal heat sources, *Energy*, 32: 1698–1706.

Mago, P., Chamra, L., Srinivasan, K. and Somayaji, C. (2008) An examination of regenerative organic Rankine cycles using dry fluids, *Applied Thermal Engineering*, 28: 998–1007.

Maizza, V. and Maizza, A. (1996) Working fluids in non-steady flows for waste energy recovery systems, *Applied Thermal Engineering*, 16: 579–590.

Maizza, V. and Maizza, A. (2001) Unconventional working fluids in organic Rankine-cycles for waste energy recovery systems, *Applied Thermal Engineering*, 21: 381–390.

Malohn, D.A. (1964) 1.5 kWe solar dynamic space power system, AIAA Paper 64–725.

Manolakos, D., Papadakis, G., Kyritsis, S. and Bouzianas, K. (2007) Experimental evaluation of an autonomous low-temperature solar Rankine cycle system for reverse osmosis desalination, *Desalination*, 203: 366–374.

Manolakos, D., Kosmadakis, G., Kyritsis, S. and Papadakis, G. (2009) Identification of behaviour and evaluation of performance of small scale, low-temperature organic Rankine cycle system coupled with a RO desalination unit, *Energy*, 34: 767–774.

Margolis, D.L. (1978) Analytical modelling of helical screw turbines for performance prediction, *Journal of Engineering for Power*, 100(3): 482–487.

Marques da Silva, P. (1989) Organic fluid mixtures as working fluids for the Trilateral Flash Cycle system, PhD thesis, City University, London.

Marsden, C. (1963) *Solvents Guide*, 2nd edn. New York: Interscience Publishers.

Marsden, C. (1966) *Matheson Gas Data Book*, 4th edn. Herst Litho Inc.

Mason, D.G. and Van Osdol, J.H. (1974) A 4000 H.P. reactor/organic Rankine cycle power system for undersea oil field application, *6th Annual Offshore Technology Conference*, Houston, TX.

McAnneny, A.W. (1962) Binary cycle gas fired turbine for gas compression service, *ASME Paper 62-GTP-17.*

McChesney, I.G. and Shannon, R.H. (1955) The use of benzene as a thermodynamic working fluid for a nuclear power plant, *ASME Paper 55-s-29.*

McGowan, J.G., Connell, J.W., Ambs, L.L. and Goss, W.P. (1973) Conceptual design of a Rankine cycle powered by the ocean temperature difference, *Proceedings 8th IECEC*, pp. 420–427.

McKhann, G.G., Coggi, J.V. and Diamond, S.D. (1970) Preliminary design of reactor power systems for the manned space base, *Proceedings 5th IECEC*, pp. 12.71–12.79.

McMahan, A.C. (2006) Design and optimization of Organic Rankine Cycle solar-thermal power plants, MSc thesis, University of Wisconsin-Madison.

Mellan, I. (1970) *Industrial Solvents Handbook*. Park Ridge, NJ: Noyes Data Corp.

Menssen, E. (1977) Screw Compressor with Involute Profiled Teeth. US Patent 4,028,026.

Mercury Power Plant from South Meadow, 1928 to Schiller, 1949 (1950) *Power Generation*, 54(3): 54–55.

Meyer, C.A. and Fischer, F.K. (1962) Working fluids for power generation of the future, *Proceedings American Power Conference*, 24: 371–380.

Meyers, K. (1997) Creating the right environment for machine bearings, *Evolution, SKF Industrial Journal*, 4: 21–24.

Mikielewicz, D. and Mikielewicz, J. (2010) A thermodynamic criterion for selection of working fluid for subcritical and supercritical domestic micro CHP, *Applied Thermal Engineering*, 30(16): 2357–2362.

Miller, D.R., Thompson, Q.E., Hull, H.R. and Barber, K.F. (1972) Working fluids for automotive Rankine engines, *Proceedings 7th IECEC*, pp. 315–326.

Misra, L.N. (1972) Prospects of power plant cycles using refrigerants, *ASHRAE Journal*, 14: 55–59.

Miura, T. and Sega, K. (1969) Centrifugal refrigeration plant driven by fluorocarbon turbine, *IHI Engineering Review*, 2(2): 12–19.

Monahan, J. (1976) Development of a 1-kW, Organic Rankine Cycle power plant for remote applications,

Intersociety Energy Conversion Engineering Conference, New York.

Morgan, D.T. and David, J.P. (1974) High efficiency gas turbine/organic Rankine cycle combined power plant, *ASME Paper 74-GT-35*.

Morgan, D.T., Doyle, E.F. and Kitrilakis, S.S. (1969) Organic Rankine cycle with reciprocating engine, *Proceedings 4th IECEC*, pp. 1–10.

Morgan, D., Raymond, R., Machacok, R. and Dawson, D. (1970) Rankine cycle power system with organic fluid and reciprocating engine for low emission automotive propulsion, *Proceedings 5th IECEC*, pp. 11.45–11.55.

Morgan, D.T., Doyle, E. and Hutchins, P. (1972) Low emission Rankine cycle engine with organic based working fluid and reciprocating expander for automobiles, *Proceedings 7th IECEC*, pp. 913–926.

Morgan, D., Patel, P., Doyle, E., Raymond, R., Sakhuja, S. and Barber, K. (1973) Laboratory tests results in low emission Rankine cycle engine with organic based working fluid and reciprocating expander for automobiles, *Proceedings 8th IECEC*, pp. 158–164.

Moss, T.A., Wilson, R.F. and Niggemann, R.E. (1970) 66 kWe ZrH reactor–organic Rankine power systems for large manned orbiting space systems, *Proceedings 5th IECEC*, pp. 12.49–12.57.

Muller-Steinhagen, H. and Trieb, F. (2004) Concentrating solar power – a review of the technology, *Quarterly of the Royal Academy of Engineering, Ingenia*, 18: 43–50.

Nedjat, Z. (1965) Waste heat recovery for a gas turbine, MSc thesis, Imperial College, University of London.

Nelson, C. (2009) Exhaust energy recovery, in *Directions in Engine Efficiency and Emissions Research Conference*,

Detroit, MI, 3–6 August. Washington, DC: US Department of Energy, ACE 041.

Nguyen, V., Doherty, P. and Riffat, S. (2001) Development of a prototype low temperature Rankine cycle electricity generation system, *Applied Thermal Engineering*, 21: 169–181.

Niggemann, R.E. (1972) 3000 hour endurance test of a 6 kWe organic Rankine cycle power system, *Proceedings 7th IECEC*, pp. 288–295.

Niggemann, R.E., Bland, T.J. and Wigmore, D.B. (1970) A 6 kWe organic Rankine power conversion system for space applications, *Proceedings 5th IECEC*, pp. 11.56–11.61.

Nilsson, H.R. (1952) US Patent 2,622,787.

Norris, T.D., Horn, G. and Sanderson, P.R. (1963) Proposals for a gas turbine-refrigerant cycle, *CEGB Report RD/M/N49*.

Norton, J.P. (1968) Technical assessment of a turbine powered heater, *Proceedings 3rd IECEC*, pp. 370–375.

Norton, J.P. (1969) Design of a practical hydrocarbon boiler, *Proceedings 4th IECEC*, pp. 21–25.

Nouri, J.M., Gueratto, D., Stosic, N., Kovacevic, A. and Arcoumanis, C. (2006) Cycle resolved velocity measurements within a screw compressor, *Proceedings International Compressor Engineering Conference*, Paper 1810. Available at: http://docs.lib.purdue.edu/icec/1810.

Nouri, J.M., Gueratto, D., Stosic, N., Smith, I.K. and Arcoumanis, D. (2007) Flow measurements in the discharge port of a screw compressor, *Journal of Process Mechanical Engineering, Part E*, 222(E4): 211–223.

Nowak, W., Borsukiewicz-Gozdur, A. and Stachel, A. (2008) Using the low-temperature Clausius–Rankine cycle to cool technical equipment, *Applied Energy*, 85: 582–588.

Ohman, H. (1999) US Patent 5,947,713.

Ohman, H. and Bishop, C.J. (1973) A look at solar energy for Seattle, *Proceedings 8th IECEC*, pp. 360–365.

O'Neill, P.A. (1993) *Industrial Compressors: Theory and Equipment.* Oxford: Butterworth- Heinemann.

Otteson, R. (1966) A survey of the isotope fuelled small Rankine cycle power unit, *UKAEA/ENEA Symposium on Industrial Applications for Isotopic Power Generators*, Harwell, September.

Owen, J.R. (1975) The Organic Rankine Cycle: a review of applications and factors affecting working fluid selection, *City University Research Memorandum ML 76*, June.

Pack, C.R. and Toft, B.P. (1969) The use of fluorocarbons for marine propulsion, *RNC Greenwich Advanced Marine Engineering Course Project Report*, July.

Papadopoulos, A., Stijepovic, M. and Linke, P. (2010) On the systematic design and selection of optimal working fluids for Organic Rankine Cycles, *Applied Thermal Engineering*, 30: 760–769.

Parisi, A.J. (1968) Great future is predicted for Oracle power systems, *Product Engineering*, 39(14): 14–15.

Patel, P.S. and Doyle, E.F. (1976) Compounding the truck diesel engine with an Organic Rankine-Cycle system, *SAE Technical Paper 760343*.

Pedro, J., Chamra, L., Srinivasan, K. and Chandramohan, S. (2008) An examination of regenerative organic Rankine cycle using dry fluids, *Applied Thermal Engineering*, 28: 998–1007.

Percival, W.H. (1965) Fluorochemical vapour machine, *SAE Transactions*, 73: 549–560.

Piacentino, A. and Cardona, F. (2007) On thermoeconomics of energy systems at variable load conditions: integrated optimisation of plant design and operation, *Energy Conversion and Management*, 48: 2341–2355.

Pitzer, K.S. and Curl, R.F., Jr. (1957) The volumetric and thermodynamic properties of fluids. III. Empirical equation for the second virial coefficient, *Journal of the American Chemical Society*, 79: 2369–2370.

Pitzer, K.S., Lippmann, D.Z., Curl, R.F., Jr., Huggins, C.M. and Petersen, D.E. (1955a) The volumetric and thermodynamic properties of fluids. I. Theoretical basis and virial coefficients, *Journal of the American Chemical Society*, 77(13): 3427–3432.

Pitzer, K.S., Lippmann, D.Z., Curl, R.F., Jr., Huggins, C.M. and Petersen, D.E. (1955b) The volumetric and thermodynamic properties of fluids. II. Compressibility factor, vapor pressure and entropy of vaporisation, *Journal of the American Chemical Society*, 77(13): 3433–3440.

Poling, B.E., Grens, E.A. and Prausnitz, J.M. (1981) Thermodynamic properties from a cubic equation of state – avoiding trivial roots and spurious derivatives, *Industrial and Enginnering Chemistry Process Design and Development*, 20(1): 127–130.

Poling, B.E., Prausnitz, J.M. and O'Connell, J.P. (2001) *The Properties of Gases and Liquids*, 5th edn. New York: McGraw-Hill.

Porter, S.R. (1973) The design and development of small high pressure ratio radial turbines, DPhil thesis, Oxford University.

Prabhu, E. (2006) Solar trough organic Rankine electricity system (stores) stage 1: power plant optimization and economics, *NREL Report SR-550-39433*.

Prigmore, D.R. and Barber, R.E. (1974) A prototype solar powered Rankine cycle system providing residential air conditioning and electricity, *Proceedings 9th IECEC*, pp. 326–333.

Prigmore, D. and Barber, R. (1975) Cooling with the sun's heat: design considerations and test data for a Rankine cycle prototype, *Solar Energy*, 17: 185–192.

Probert, S.D., Hussein, M., O'Callaghan, P.W. and Bala, E. (1983) Design optimisation of a solar-energy harnessing system for stimulating an irrigation pump, *Applied Energy*, 15(4): 299–321.

Purdy, D.C. (1957) Diphenyl as a thermodynamic fluid, *Nucleonics*, 15(4): 109–112.

Quoilin, S. (2011) Sustainable energy conversion through the use of organic Rankine cycles for waste heat recovery and solar applications, PhD thesis, University of Liege.

Quoilin, S. and Lemort, V. (2009) Technological and economical survey of Organic Rankine Cycle systems, *European Conference on Economics and Management of Energy in Industry*, Algarve, Portugal.

Quoilin, S., Orosz, M. and Lemort, V. (2008) Modeling and experimental investigation of an organic Rankine cycle using scroll expander for small solar applications, *Eurosun Conference*, Lisbon, Portugal.

Radwan, M.S. (1974) Internal combustion engine/vapour cycle combination, *Oxford University Engineering Laboratory Report 1097/74*.

Ray, S.K. and Moss, G. (1966) Fluorochemical as working fluid for small Rankine cycle power units, *Advanced Energy Conversion*, 6: 89–102.

Redlich, O. and Kwong, J.N.S. (1949) On the thermodynamics of solutions. V. An equation of state. Fugacities of gaseous solutions, *Chemical Reviews*, 44: 233–244.

Reid, R.C. and Sherwood, T.K. (1966) *The Properties of Gases and Liquids*, 2nd edn. New York: McGraw-Hill.

Reid, R.C., Prausnitz, J.M. and Poling, B.E. (1987) *The Properties of Gases and Liquids*, 4th edn. New York: McGraw-Hill.

Renner, R.A. (1973) Reviewing the California steam bus project, *SAE Transactions*, 81(2): 838–851.

Rentizelas, A., Karellas, S., Kakaras, E. and Tatsiopoulos, I. (2009) Comparative techno-economic analysis of ORC and gasification for bioenergy applications, *Energy Conversion and Management*, 50(3): 674–681.

Reti, G.R. (1965) Thermodynamic analysis of a new combined cycle, *Proceedings American Power Conference*, 27: 521–529.

Richter, H.R.D. and Burnside, B.M. (1975) A general program for producing pressure–enthalpy diagrams, *Journal of Mechanical Engineering Science*, 17(1): 31–39.

Rinder, L. (1979) *Schraubenverdichter* (*Screw Machines*). Berlin: Springer-Verlag [in German].

Rinder, L. (1987) Screw Rotor Profile and Method for Generating. US Patent 4,643,654.

Robert, N. (1952) Helical Rotary Engine. US Patent 2,622,787.

Rogers, G.F.C. and Mayhew, Y.R. (1959) *Engineering Thermodynamics, Work and Heat Transfer*. London: Longmans.

Roy, J.P., Mishra, M.K. and Misra, A. (2011) Performance analysis of an Organic Rankine Cycle with superheating under different heat source temperature conditions, *Applied Energy*, 88(9): 2995–3004.

Rozenfel'd, L.M. and Vorob'ev, I.D. (1969) Calculation of direct thermodynamic Freon 12 cycles using digital computers, *Thermal Energy*, 16(2): 87–91. [Translation from *Teploenergetika*, 16(2): 59–62.]

Saitoh, T., Yamada, N. and Wakashima, S. (2007) Solar Rankine cycle system using scroll expander, *Journal of Energy and Engineering*, 2: 708–718.

Sakun, I.A. (1960) *Vintoviekompresorii*. Leningrad: Mashinostroenie [in Russian].

Saleh, B., Koglbauer, G., Wendland, M. and Fischer, J. (2007) Working fluids for low temperature organic Rankine cycles, *Energy*, 32: 1210–1221.

Sangfors, B. (1982) Analytical modeling of helical screw compressor for analysis and performance prediction, *Proceedings International Compressor Engineering Conference*, Paper 384. Available at: http://docs.lib.purdue.edu/icec/384.

Sangfors, B. (1984) Computer simulation of the oil injected twin screw compressor, *Proceedings International Compressor Engineering Conference*, Paper 502. Available at: http://docs.lib.purdue.edu/icec/502.

Sargent, S.L. and Teagan, W.P. (1973) Compressor refrigeration for a solar powered organic Rankine cycle engine, *ASME Paper 73-WA-Sol 8*.

Sauls, J. (1994) The influence of leakage on the performance of refrigerant screw compressors, *Proceedings VDI Tagung 'Schraubenmaschinen 94', Dortmund VDI Berichte 1135*.

Sauret, E. and Rowlands, A.S. (2011) Candidate radial-inflow turbines and high-density working fluids for geothermal power systems, *Energy*, 36(7): 4460–4467.

Scheflan, L. and Jacob, M.B. (1953) *The Handbook of Solvents*. London: Macmillan.

Schibbye, C.B. (1979) Screw-Rotor Compressor with Straight Flank Sections. US Patent 4,140,445.

Schuster, A., Karl, J. and Karellas, S. (2007) Simulation of an innovative stand-alone solar desalination system using an organic Rankine cycle, *International Journal of Thermodynamics*, 10: 155–163.

Sega, K. (1974) Freon turbine, *Chemical Economy and Engineering Review*, 6(8): 27–35.

Shcheglayer, A.V. and Pakshuer, V.B. (1968) Thermal power stations, *Transactions 7th World Power Conference*, Moscow.

Shinbrot, C.H. and Coggi, J.V. (1969) Electrical/thermal integration of dynamic power and life support systems for manned spacecraft, *Proceedings 4th IECEC*, pp. 505–514.

Silver, R.S. (1969) Some aspects of the development of the steam condenser from the time of James Watt, *Proceedings Symposium to Commemorate the Bicentenary of the James Watt Patent*, University of Glasgow, September.

Simmons, L.D. (1972) Optimisation of staged Rankine energy conversion cycles for high efficiency, *Proceedings 7th IECEC*, pp. 279–287.

Singh, P.J. and Patel, G.C. (1984) A generalized performance computer program for oil flooded twin-screw compressors, *Proceedings International Compressor Engineering Conference*, Paper 504. Available at: http://docs.lib.purdue.edu/icec/504.

Smith, I.K. (1977) A program to determine the thermodynamic properties of organic fluids in the saturated, superheated and supercritical states by means of the Lee Kesler equations of state, *The City University, Research Memorandum ML 98*, July.

Smith, I.K. (1978) The use of the Lee-Kesler equations of state for the estimation of common refrigerant properties, *Proceedings NPL Conference on Chemical Thermodynamic Data on Fluids and Fluid Mixtures*, National Physical Laboratory, Teddington, September.

Smith, I.K. (1981) The choice of working fluids for power recovery from waste heat streams, Transactions of the Institute of Marine Engineers Conference on Organic Fluids for Waste Heat Recovery in Ships and Industry, pp. 8–18, January.

Smith, I.K. (1984) THERPROP: a computer program and subprogram set with data bank for volatile fluid thermal property estimation, City University, London, March.

Smith, I.K. (1992) Matching and work ratio in elementary power plant theory, *Proceedings IMechE, Part A*, 206(A4): 257–262.

Smith, I.K. (1993) Development of the trilateral flash cycle system. Part 1: Fundamental considerations, *Proceedings IMechE, Part A*, 207(A3): 179–194.

Smith, I.K. (1994) Existing installations, in A.W. Crook (ed.) *Profiting from Low Grade Heat: Thermodynamic Cycles for Low-Temperature Heat Sources*, pp. 127–149. London: The Institution of Electrical Engineers.

Smith, I.K. and Pitanga Marques, R. (1993) Development of the trilateral flash cycle system. Part 2: Increasing power output with working fluid mixtures, *Proceedings IMechE, Part A*, 207(A3): 179–194.

Smith, I.K. and Stosic, N. (1995) The Expressor: an efficiency boost to vapour compression systems by power recovery from the throttling process, in *AES, Vol. 34: Heat Pump and Refrigeration Systems Design*, pp. 173–178. Edinburgh: ASME.

Smith, I.K. and Stosic, N. (1999) Expanding a Liquid to its Vapour, Patent GB 2,309,748; US Patent 5,833,445.

Smith, I.K. and Stosic, N. (2001) Prospects for energy conversion efficiency improvements by the use of two-phase expanders, *Proceedings of the 2nd International Heat Powered Cycles Conference*, Paris.

Smith, I.K. and Stosic, N. (2008) Power Recovery from Exhaust Gases, Patent Application GB 0802315.2.

Smith, I.K., Stosic, N. and Aldis, C.A. (1993) Development of the trilateral flash cycle system. Part 3: The design of high efficiency screw expanders, *Proceedings IMechE, Part A*, 207(A3): 179–194.

Smith, I.K., Stosic, N. and Kovacevic, A. (2004) Power recovery from low-cost two-phase expanders, *Proceedings Geothermal Resources Council Annual Meeting*, Indian Wells, CA, August.

Smith, I.K., Stosic, N. and Kovacevic, A. (2005a) Expander Process Fluid Lubrication, Patent Application GB 0526413.0.

Smith, I.K., Stosic, N. and Kovacevic, A. (2005b) Screw expanders increase output and decrease the cost of geothermal binary power plant Systems, *Proceedings Geothermal Resources Council Annual Meeting*, Reno, NV, September.

Smith, I.K., Stosic, N., Kovacevic, A. and Mujic, E. (2011) Steam as the working fluid for power recovery from exhaust gases by means of screw expanders, *Proceedings International Conference on Compressors and Their Systems*, IMechE, London, September.

Soave, G. (1972) Equilibrium constants from a modified Redlich-Kwong equation of state, *Chemical Engineering Science*, 27(6): 1197–1203.

Solberg, H.L. and Hawks, K.H. (1971) The thermal performance of fluids in a Rankine cycle power plant for automotive applications, *Proceedings 6th IECEC*, pp. 813–823.

Somayaji, C., Mago, P. and Chamra, L. (2006) Second law analysis and optimization of organic Rankine cycles, *ASME Power Conference Paper PWR2006-88061*, Atlanta, GA.

Southam, D.L. (1964) Power systems comparison for manned space applications, in *Progress in Astronautics and Aeronautics, Vol. 16: Space Power Systems Engineering*, pp. 335–359.

Sprankle, R. (1973) Electrical Power Generating Systems. US Patent 3,751,673.

Starr, M.D. and Hagey, G.L. (1968) A Navy 2–10 kWe radioisotope power system for undersea applications, *Proceedings 3rd IECEC*, pp. 805–810.

Steidel, R. and Berger, R. (1981) Performance characteristics of the Lyshlom engine as tested for geothermal applications, *Proceedings International Energy Conversion Engineering Conference*, pp. 1334–1340.

Steigelmann, W.H., Seth, R.G. and Wachtell, G.P. (1972) Binary cycle power plants using air-cooled condensing systems, *Proceedings American Power Conference*, 34: 521–530.

Stoddart, D.E. (1968) Waste heat recovery by the use of low boiling point liquids, *Process Engineering*, November, pp. 5–8.

Stosic, N. (1996) Patent Application GB 9610289.2.

Stosic, N. (1998) On gearing of helical screw compressor rotors, *Proceedings IMechE, Journal of Mechanical Engineering Science*, 212: 587–594.

Stosic, N. (2004) Screw compressors in refrigeration and air conditioning, *International Journal of HVACR Research*, 10(3): 233–263.

Stosic, N. (2006) A geometric approach to calculating tool wear in screw rotor machining, *International Journal of Compressor Tools and Manufacture*, 46(15): 1961–1965.

Stosic, N. and Hanjalic, K. (1997) Development and optimization of screw compressors with a simulation model. Part I. Profile generation, *ASME Transactions, Journal of Fluids Engineering*, 119: 659.

Stosic, N., Milutinovic, Lj., Hanjalic, K. and Kovacevic, A. (1992) Investigation of the influence of oil injection upon the screw compressor working process, *International Journal of Refrigeration*, 15(4): 206–219.

Stosic, N., Smith, I.K., Kovacevic, A. and Aldis, C.A. (1997) The design of a twin-screw compressor based on a new profile, *Journal of Engineering Design*, 8: 389–399.

Stosic, N., Smith, I.K. and Kovacevic, A. (2002) A twin screw combined compressor and expander for CO_2 refrigeration systems, *Proceedings International Compressor Engineering Conference*, Paper 1591. Available at: http://docs.lib.purdue.edu/icec/1591.

Stosic, N., Smith, I.K. and Kovacevic, A. (2003a) Opportunities for innovation with screw machines, *Proceedings IMechE, Journal of Process Mechanical Engineering*, 217: 157–170.

Stosic, N., Smith, I.K. and Kovacevic, A. (2003b) Rotor interference as a criterion for screw compressors design, *Journal of Engineering Design*, 14(2): 209–220.

Stosic, N., Smith, I.K. and Kovacevic, A. (2003c) Optimisation of screw compressors, *Applied Thermal Engineering*, 23(10): 1177–1195.

Stosic, N., Smith, I.K. and Kovacevic, A. (2005) *Screw Compressors, Mathematical Modeling and Performance Calculation*. Berlin: Springer-Verlag.

Stosic, N., Smith, I.K., Kovacevic, A. and Mujic, E. (2010) Research in water lubricated screw machines for fuel cell application, *FC Expo, Academic Forum*, Tokyo, March.

Tabor, H. (1962) Solar utilisation and development in Israel, *Transactions 6th World Power Conference*, 8: 3120–3132.

Tabor, H.Z. (1967) Power for remote areas, *International Science and Technology*, 65: 52–59.

Tabor, H. and Bronicki, L. (1961) Small turbine for solar energy power package, *Proceedings UN Conference on New Sources of Energy*, 4: 68–79.

Tabor, H. and Bronicki, L. (1965) Establishing criteria for fluids for small vapour turbines, *SAE Transactions*, 73: 561–575.

Tang, Y. and Fleming, J.S. (1992) Obtaining the optimum geometrical parameters of a refrigeration helical screw compressor, *Proceedings International Compressor*

Engineering Conference, Paper 811. Available at: http://docs.lib.purdue.edu/icec/811.

Tang, Y. and Fleming, J.S. (1994) Clearances between the rotors of helical screw compressors: their determination, optimisation and thermodynamic consequences, *Journal of Process Mechanical Engineering*, 208: 155–163.

Taniguchi, H., Kudo, K., Giedt, W.H., Park, I. and Kumazawa, S. (1988) Analytical and experimental investigation of two-phase flow screw expanders for power generation, *Transactions ASME, Journal of Engineering for Gas Turbines and Power*, 110: 628–635.

Teagen, W.P. (1972) Rankine cycle power in-plant vehicles, *Proceedings 7th IECEC*, pp. 905–912.

Thurston, R.H. (1902) Series vapour and waste heat engines as supplementary to single vapour engines, *Journal of the Franklin Institute*, 154(4): 291–314; 154(5): 367–390; 154(6): 447–479.

Timmermans, J. (1950) *Physico-chemical Constants of Pure Organic Compounds*, Vol. 1. New York: Elsevier.

Timmermans, J. (1965) *Physico-chemical Constants of Pure Organic Compounds*, Vol. 2. New York: Elsevier.

Tsederberg, N.V. (1965) *Thermal Conductivity of Gases and Liquids*. London: Arnold.

Van Osdol, J.H., Wilson, R.F. and Niggemann, R.E. (1969) Zirconium hydride reactor-organic Rankine power systems, *Proceedings 4th IECEC*, pp. 456–464.

Van Osdol, J.H., Hengle, J.E., Reardon, D.E. and Wilson, R.F. (1972) Nuclear power system study, *Proceedings 7th IECEC*, pp. 559–567.

Vijayaraghavan, S. and Goswami, D. (2005) Organic working fluids for a combined power and cooling cycle, *Journal of Energy Resources Technology*, 127: 125–130.

Vild, T.J., Schubert, F.H., Snoke, D.R. and Delaney, C.L. (1968) A system to demonstrate the zero gravity performance of an organic Rankine cycle, *Proceedings 3rd IECEC*, pp. 376–388.

Walas, S.M. (1985) *Phase Equilibria in Chemical Engineering*. Stoneham, MA: Butterworths.

Wang, X., Zhao, L., Wang, J., Zhang, W., Zhao, X. and Wu, W. (2010) Performance evaluation of a low-temperature solar Rankine cycle system utilizing R245fa, *Solar Energy*, 84: 353–364.

Weast, R. (ed.) (1974/75) *CRC Handbook of Chemistry and Physics*, 55th edn. Boca Raton, FL: CRC Press.

Wei, D., Lu, X., Lu, Z. and Gu, J. (2007) Performance analysis and optimization of organic Rankine cycle for waste heat recovery, *Energy Conversion and Management*, 4: 113–119.

Wei, D., Lu, X., Lu, Z. and Gu, J. (2008) Dynamic modeling and simulation of an Organic Rankine Cycle (ORC) system for waste heat recovery, *Applied Thermal Engineering*, 28(10): 1216–1224.

Werner, D.K. and Barker, R.E. (1973) Working fluid selection for a small Rankine cycle total energy system for recreation vehicles, *Proceedings 8th IECEC*, pp. 146–151.

Wigmore, D.B., Niggemann, R.E. and O'Sullivan, J.B. (1972) The specification of an optimum working fluid for a small Rankine cycle turboelectric power system, *Proceedings 7th IECEC*, pp. 303–314.

Wilson, S.S. (1964) *The Rankine Power Pack*. Oxford: Oxford University Engineering Laboratory.

Wilson, S.S. (1967) New working fluids for power units, *New Scientist*, 36(571): 412–413.

Wilson, S.S. and Radwan, M.S. (1977) Appropriate thermodynamics for heat engine analysis and design, *International Journal of Mechanical Engineering Education*, 5(1).

Wolpert, J.L. and Riffat, S.B. (1996) Solar-powered Rankine system for domestic applications, *Applied Thermal Engineering*, 16: 281–289.

Wood, B. (1969/70) Alternative fluids for power generation, *Proceeding IMechE*, 184 (Pt. 1): 713–740.

Wood, De V. (1891) Theoretical investigation of the efficiency of vapour engines, *Transactions ASME*, 12: 155–173.

Xiaojun, G., Liansheng, L., Yuanyang, Z., Pengcheng, S. and Shen, J. (2010) Research on a scroll expander used for recovering work in a fuel cell, *International Journal of Thermodynamics*, 7(1): 1–8.

Xing, Z.W. (2000) *Screw Compressors*. Beijing: Machine Press.

Xion, Z. and Dagang, X. (1986) Study on actual profile surface and engaging clearance of screw compressor rotors, *Purdue Compressor Technology Conference Paper 239*.

Yagoub, W., Doherty, P. and Riffat, S. (2006) Solar energy-gas driven micro-CHP system for an office, *Applied Thermal Engineering*, 26: 1604–1610.

Yamadaa, N., Minamib, T. and Mohamadb, A. (2011) Efficiency of compact Organic Rankine Cycle system with rotary-vane-type expander for low-temperature waste heat recovery, *Energy*, 36: 1010–1017.

Yamamoto, T., Furuhata, T., Arai, N. and Mori, K. (2001) Design and testing of the organic Rankine cycle, *Energy*, 26: 239–251.

Yanagisawa, T., Fukuta, Y., Ogi, T. and Hikichi, T. (2001) Performance of an oil-free scroll-type air expander, *International Conference on Compressors and Their Systems*, pp. 167–174.

Yarrow, A.F. (1888) On the possible advantage of using highly volatile liquids in lieu of water for the purposes of propulsion, *Transactions Institute of Naval Architects*, 29: 248–256.

Yellott, J.I. (1962) Solar energy progress report – 1961, *Journal of Engineering for Power*, 84(2): 213–221.
Yellott, J.I. (1964) Solar energy progress report – 1963, *Journal of Engineering for Power*, 86(4): 475–484.
Yellott, J.I. (1966) Solar energy progress report – 1965, *Journal of Engineering for Power*, 88(3): 221–231.
Ziervogel, R.G. and Poling, B.E. (1983) A simple method for constructing phase envelopes for multicomponent mixtures, *Fluid Phase Equilibria*, 11(2): 127–135.
Zwick, E.B. (1970) The smog free engine of tomorrow – today, Proceedings 5th IECEC, pp. 17.19–17.28.

Index

FSC
www.fsc.org
MIX
Papier | Fördert gute Waldnutzung
FSC® C083411

Kolibri 360 GmbH
Ferdinand-Jühlke-Straße 7
99095 Erfurt, Deutschland
produktsicherheit@kolibri360.de